Systems with Small Dissipation

T0138266

Systems with Small Dissipation

V. B. Braginsky
V. P. Mitrofanov
V. I. Panov

Translated by Erast Gliner
Edited by Kip S. Thorne and Cynthia Eller

The University of Chicago Press
Chicago and London

V. B. Braginsky is professor of physics and Chair of the Division of Radiophysics and Electronics of the Department of Physics at Moscow State University. He is also Chair of the Gravitation Commission of the Academy of Sciences of the USSR.

V. P. Mitrofanov and V. I. Panov are senior scientists at Moscow State University.

Erast Gliner is research associate at the McDonnell Center for the Space Sciences, Washington University, St. Louis.

Kip S. Thorne is the William R. Kenan, Jr., Professor and professor of theoretical physics at the California Institute of Technology.

Cynthia Eller is administrative secretary at the California Institute of Technology and a Ph. D. candidate in social ethics at the University of Southern California.

The University of Chicago Press, Chicago 60637
The University of Chicago Press, Ltd., London

Library of Congress Cataloging-in-Publication Data

Braginskiĭ, V. B. (Vladimir Borisovich)
    Systems with small dissipation.

    Translation of: Sistemy s maloĭ dissipatsieĭ.
    Bibliography: p.
    Includes index.
    1. Physical measurements. 2. Harmonic oscillators--
Design and constuction. 3. Resonators--Design and
construction. I. Mitrofanov, V. B. II. Panov, V. I.
(Vladimir Ivanovich) III. Thorne, Kip S. IV. Eller,
Cynthia. V. Title. VI. Title: Dissipation.
QC39.B66313 1985   530.8   85-20876
ISBN 0-226-07072-7
ISBN 0-226-07073-5 (pbk.)

Originally published as *Sistemi s Maloi Dissipatsiei* in Moscow in 1981 by Nauka. This English-language edition is published through an agreement with the Copyright Agency of the Union of Soviet Socialist Republics (VAAP).

Contents

Foreword

For five years in the late 1970s and early 1980s, when my own research was focussing on the theory of high-precision measurements, my good friend Murray Gell-Mann took delight in asking me, "Well, Kip, have you finished solving the simple harmonic oscillator yet?" My invariable reply was, "No, not yet; I'm still struggling with it!"

To a student of elementary physics I might have seemed a fool. But as Gell-Mann and I, sharing our own private joke, were well aware, the simple harmonic oscillator is the key tool that permits experimenters to detect extremely weak mechanical forces and electromagnetic signals and to produce highly stable standards of time and frequency. The oscillator, for example, underlies radio and microwave receivers, gravitational-wave detectors, clocks, searches for quarks, tests of the equivalence principle, and tests of theories of superfluidity and superconductivity.

The central problem of such experiments is to construct an oscillator that is as perfectly simple harmonic as possible, and the largest obstacle to such construction is the oscillator's dissipation. If dissipation were perfectly smooth, it would not be much of an obstacle, but the fluctuation-dissipation theorem of statistical mechanics guarantees that any dissipation is accompanied by fluctuating forces. The stronger the dissipation, the larger the fluctuating forces, and the more seriously they mask the signals that the experimenter seeks to detect.

This book is a treatise on the sources of dissipation and other defects in mechanical and electromagnetic oscillators and on practical techniques for minimizing them. The viewpoint of the book is that of an experimenter who wants to construct the most nearly perfect measuring apparatus possible, and who needs information about everything from the role of phonon-photon scattering as a fundamental source of dissipation to the effectiveness of a thin film of pork fat in reducing the friction between a support wire and a mechanically oscillating sapphire crystal.

The authors are among the world's outstanding experts on oscillators with small dissipation. Since the mid 1960s Vladimir Braginsky has led a research team at Moscow State University that focuses on the measurement of extremely weak forces and small signals. V. P. Mitrofanov and V. I. Panov are two of the most senior and most talented experimenters on Braginsky's team. The members of the Braginsky team are best known in the West for their contributions to the technology of gravitational-wave detection, their experimental search for quarks, their test of the equivalence principle, and their invention of new experimental techniques for high-precision measurement including "quantum nondemolition measurements." Their ingenuity and experimental talent have enabled them to compete with experimenters in the West and, despite the inferiority of the Soviet technological base, to win that competition more often than not.

In the process of their research, the members of the Braginsky team have acquired deep insight into the practical aspects of constructing oscillators with very low dissipation. This book presents their accumulated wisdom and is thus a treasure trove of useful facts, viewpoints, and tricks.

It may seem peculiar that I, a theoretical physicist, would edit the English translation of this book. However, I have done it for selfish reasons: Much of my current research entails thinking about high-precision measurements, albeit from a far less practical viewpoint than that of this book, and I believed (correctly) that while editing the book I would learn a lot that would help me in my work. In addition, I find this subject fascinating, and I have long found Braginsky and his team fascinating. The price of my editing the book, however, is an apology to the reader: since I am not an expert in experimental physics, much less in practical aspects of low-dissipation oscillators, I probably have mutilated the correct English technological phraseology here and there.

The mutilations would have been far worse if it were not for the care with which Erast Gliner, the translator (and also an excellent physicist), performed the translation. And my task would have been far harder if it were not for the crucial role played by my co-editor, Cynthia Eller, who converted Gliner's translation into colloquial English without distorting the technical meanings (Russian, not English, is Gliner's native language). While editing her text I referred not only to Gliner's original translation but also to the Russian original. After the editing was finished, Braginsky read through the text for technical accuracy, and several other experimentalist colleagues (David Blair, John Dick, Ron Drever, Bob Spero, Bob Vessot, and Stan Whitcomb) read portions of it and suggested changes, for which I thank them. This makes me hopeful that the only serious errors are those of occasionally unconventional technical phraseology. For such errors, and any others, I must bear the greatest blame.

Kip S. Thorne

Preface to the English Edition

This book has been written by experimentalists and for experimentalists. Our main goal has been to join together in one place all information about the present state of the art in the construction and performance of high-Q mechanical and electromagnetic resonators, about the fundamental and practical obstacles that limit their levels of losses, and about some possible applications of such resonators in modern physics experiments.

Since the Russian edition of this book was published in 1981, there have been no significant new techniques proposed for the design and construction of resonators with low levels of losses. However, significant new progress has been made on practical aspects of achieving low losses, and this English edition has been updated to include information about that progress.

It is a great pleasure for us, the authors, to express our sincere thanks to the scientific editors, Academician Kip S. Thorne and Ms. Cynthia Eller, for their scrupulous attention to the English edition, and to our colleagues R. W. P. Drever, S. E. Whitcomb, G. J. Dick, and D. Strayer for their critical remarks.

V. B. Braginsky
V. P. Mitrofanov
V. I. Panov

Preface to the Russian Edition

Mechanical and electromagnetic resonators with low levels of losses are used widely throughout experimental physics and engineering. An increasing number of experimental studies use high-Q circuits or mechanical oscillators as key elements. For example, macroscopic oscillators with small losses are crucial components in secondary frequency standards and in second-generation gravitational-wave antennas, and they are used widely in radiospectroscopic experiments that measure small changes in the properties of semiconducting and dielectric materials at very high frequencies. This extensive use of high-Q systems is connected on the one hand with their power and simplicity as tools for measuring many physical quantities and on the other hand with continuing refinements in the art of constructing high-Q systems.

A rather long time has passed since publication of the monographs on high-Q systems by A. G. Smagin and M. I. Yaroslavsky (1970) and A. I. Didenko (1973). During this period, a large number of research papers have addressed either methods of obtaining low levels of dissipation in macroscopic oscillating systems or applications of those systems in physics experiments.

We have often used high-Q oscillators of various types in our laboratory, and we have written this book based on our own experience and our studies of the work of others. Included in this review are short descriptions of present viewpoints about various physical mechanisms of energy dissipation in macroscopic oscillating systems, descriptions of many technical procedures and methods leading to high quality factors, and also short descriptions of several types of important physical experiments in which the key element is a high-Q oscillator.

Recent theoretical investigations have revealed that when changes in the level of oscillation are very small, a macroscopic oscillator must be regarded as a quantum system even at fairly high temperatures. This conclusion plays a central role in experiments where it is necessary, for example, to detect the effects of a very weak force that produces an amplitude change smaller than the width of a wave packet in a coherent quantum state. Theoretical studies have also revealed that when the addition or removal of energy from a macroscopic oscillator is measured, its behavior will be quantum mechanical (the oscillator will have a discrete set of energy levels) if its Q is sufficiently great. This is true even if the mean thermal energy in the oscillator is considerably larger than the distance between quantum levels. This book includes a short summary of this new branch of the "quantum theory of measurement" and a discussion of its possible practical applications.

V. B. Braginsky
V. P. Mitrofanov
V. I. Panov

I   Introduction

1. Classical Oscillators with Small Dissipation

Dissipation in a linear oscillator is commonly described by the relaxation time τ^*, or, equivalently, the dissipation rate $\delta = (\tau^*)^{-1}$, or by either of two dimensionless quantities, the quality factor Q and the relaxation factor θ. These quantities are related to the oscillator's physical parameters as follows: The free oscillation equation for an oscillator with mass M, spring constant (rigidity) K, and friction coefficient H has the form

$$M\ddot{x} + H\dot{x} + Kx = 0 , \tag{1.1}$$

and has the solution

$$x(t) = x_0 \exp(-\delta t)\cos(\omega t + \varphi) , \tag{1.2}$$

where

$$\delta = \frac{H}{2M} , \quad \omega^2 = \frac{K}{M} - \delta^2 , \quad \tau^* = \delta^{-1} ,$$

$$Q = \frac{\omega\tau^*}{2} , \quad \theta = \frac{\pi}{Q} , \tag{1.3}$$

and x_0 and φ are determined by the oscillator's initial conditions.

Similar equations and expressions for δ, τ^*, Q, and θ may be written for a lumped-element circuit and for each discrete oscillation mode of a distributed-element mechanical system or circuit. The well-known basic features of any classical, underdamped

1

oscillator, as illustrated by a mechanical oscillator, will be summarized in this section.

In a heat bath at temperature T, the mean energy of each heat-bath mode is $\overline{\mathcal{E}} = kT$ (k is Boltzmann's constant). Therefore, when a mechanical oscillator is coupled to such a bath, its mean energy is

$$\overline{\mathcal{E}} = \frac{M\overline{\dot{x}^2}}{2} + \frac{K\overline{x^2}}{2} = kT .\qquad(1.4)$$

The characteristic timescale for fluctuations of the oscillator's energy is $\tau^*/2 = M/H$. To derive equation (1.4), it is sufficient to place on the right side of equation (1.1) the random force of the heat bath on the oscillator, which has a frequency-independent spectral density of $4\ kTH$ (Nyquist's fluctuation-dissipation theorem).

If the oscillator is driven not only by the heat bath but also by a force $F(t)$, which oscillates roughly at the oscillator's frequency but acts only for a finite time, then to be detected the force must have an amplitude exceeding

$$F_{\min} \geq \sqrt{4\ kTH\ \Delta f} = \left[\frac{8\ kTM\ \Delta f}{\tau^*}\right]^{\frac{1}{2}} .\qquad(1.5)$$

Here Δf is the frequency bandwidth in which the main part of the force's spectral density falls. Equation (1.5) shows unambiguously that for a classical oscillator the minimum detectable force decreases with decreasing H (or equivalently, with increasing τ^* or Q). Equation (1.5) is valid even if the duration of the force's action, $\hat{\tau}$, is much less than the relaxation time τ^*.

If $\hat{\tau}/\tau^* \ll 1$, one way to monitor the oscillator's response to the force $F(t)$ is to measure changes in its amplitude and phase. If $\hat{\tau}/\tau^* \ll 1$, the mean thermally induced amplitude change is

$$\Delta x_T \approx \left[\frac{2\ kT\hat{\tau}}{M\omega^2\tau^*}\right]^{\frac{1}{2}} = \left[\frac{kT\hat{\tau}}{M\omega Q}\right]^{\frac{1}{2}} .\qquad(1.6)$$

It is very important to note that for large τ^* ($\tau^* \gg \hat{\tau}$) and thus for large Q, the change Δx_T is far less than the root-mean-square (rms) amplitude ($\sqrt{2kT/M\omega^2}$). This permits the experimenter to monitor increases or decreases of oscillator energy that are substantially less than the equilibrium value kT.

A detailed consideration of the monitoring of weak forces by measuring an oscillator's response is outside the scope of this text (see Braginsky and Manukin 1974 for a thorough exposition). But it should be emphasized that equations (1.4) to (1.6) are valid only if the experimenter uses a position-measuring device that produces a negligible back-action force on the oscillator. Actually, any real measuring device produces some back-action force which has a dynamically fluctuating part. Even under conditions of optimal coupling, compensation of the dynamical back action, and infinite τ^* and Q, the back action gives rise to a nonnegligible, minimum measurable force F_{min}. For example, for a variable-capacitance electronic transducer, back action produces

$$F_{min} \approx \frac{2}{\hat{\tau}} \left[\frac{kT_e M\omega}{\omega_e} \right]^{1/2}, \tag{1.7}$$

where T_e is the noise temperature and ω_e is the angular frequency of the electrical circuit (Braginsky and Manukin 1974).

Under optimal coupling, equation (1.7) holds for superconducting quantum interference devices (SQUIDs), which are often used as sensors for small oscillations (Gusev and Rudenko 1976). We note in passing that if an experimenter has an amplifier with a noise temperature less than T_e, it can be used to further reduce F_{min} (Gusev and Rudenko 1978). The range of validity of equations (1.5) to (1.7) will be discussed in the next section, which deals with quantum mechanical aspects of macroscopic oscillators.

Equations (1.5) to (1.7), which are valid in the classical regime, reveal that devices for measuring weak forces have large reserves of sensitivity. Taking the values $\tau^* > 6 \times 10^7$ sec and $Q \approx 5 \times 10^9$, which have already been achieved, and assuming that $kT_e/\omega_e > \hbar$ [which is required for the validity of equation (1.7)], one can see that the actual sensitivities of nearly all existing force-measuring devices is less, by many orders of magnitude, than their limiting sensitivities. Only resonant-bar gravitational-wave detectors are now approaching these limits.

In summary, the weaker the coupling between the oscillator and the heat bath (i.e., the larger τ^*), the smaller the forces that can be detected.

A low level of oscillator damping is important not only in measurements of weak forces that act on a test body but also in many other situations, experimental techniques, and experimental apparatus where a key role is played by high-Q oscillations. Here we will mention only two examples: First, high-Q oscillators (electromagnetic and mechanical resonators) are widely used as narrow-band filters; the bandwidth of the filter is approximately equal to ω/Q and thus decreases as the damping decreases. Second, high quality factors give fairly simple ways for solving a set of technical problems associated with constructing highly stable secondary frequency standards.

In chapters 2 and 3 we will discuss the physical effects that limit the quality factors that can be achieved for various mechanical and electromagnetic systems. In chapters 4 and 5 we will briefly describe a number of experimental techniques and devices in which the weakness of an oscillator's dissipation plays an important role.

2. Quantum Mechanical Features of Macroscopic Oscillators

As already mentioned, if the "averaging time" $\hat{\tau}$ (the time between completion of measurements) is much shorter than the relaxation time τ^*, one can monitor changes of the oscillator's amplitude that are much smaller than the rms amplitude of thermal motions. Thus an experimenter can detect increases or decreases ($\Delta\mathcal{E}$) in the oscillator's energy that are considerably less than the equilibrium value kT. The smallest detectable value of $\Delta\mathcal{E}$ ($\Delta\mathcal{E}_{min}$) depends on the initial amplitude x_0 of the vibrations (at time $\tau = 0$). For instance

$$\Delta\mathcal{E}_{min} \approx 2\,kT\,\frac{\hat{\tau}}{\tau^*} \qquad \text{if} \qquad x_0 = 0\;,$$

$$\Delta\mathcal{E}_{min} \approx 2\,kT\left(\frac{\hat{\tau}}{\tau^*}\right)^{1/2} \qquad \text{if} \qquad x_0 = \left(\frac{kT}{M\omega^2}\right)^{1/2}. \qquad (2.1)$$

Completely analogous expressions can be written for a high-Q electromagnetic resonator. It is clear that a quantum mechanical analysis of the oscillator's behavior is required if $\Delta\mathcal{E}_{min} \approx \hbar\omega$.

Energy changes for a quantum oscillator in contact with a heat bath of temperature T have been analyzed by Braginsky and Nazarenko (1969). The main results of these studies are the following. If the oscillator initially is in an energy eigenstate with n quanta, then there is a large probability that it will remain in this state for a time $\hat{\tau}$, satisfying

$$(n + \tfrac{1}{2}) \, kT \, \frac{\hat{\tau}}{\tau^*} \; \ll \; \hbar\omega . \qquad (2.2)$$

The average time the oscillator remains in state n is

$$\hat{\tau}_n \approx \frac{\hbar\omega}{kT} \, \frac{\tau^*}{n} \; - \; \frac{\tau^*}{n_T^2} \quad \text{if} \quad n - n_T \equiv \frac{kT}{\hbar\omega} . \qquad (2.3)$$

Thus, in principle, an observer studying the Brownian motion of an oscillator with only the help of an energy-measuring device can detect "energy steps" equal to $\hbar\omega$, even if $kT \gg \hbar\omega$, so long as the relaxation time τ^* (the quality factor Q) is sufficiently large. If the mean time $\hat{\tau}_n$ that the oscillator stays in an energy level of given n is approximately equal to the oscillator's period, then this specific energy level can be regarded as a "threshold level" n_{th}: one can observe the discreteness of energy levels if $n < n_{th}$, but the levels begin to blur and the oscillator becomes fully classical when $n > n_{th}$.

The threshold level n_{th} is given by

$$(n_{th} + \tfrac{1}{2}) \, \frac{kT}{Q} \; \approx \; \frac{\hbar\omega}{4\pi} . \qquad (2.4)$$

For a superconducting microwave-frequency electromagnetic resonator (circuits or cavities with $\omega_e \sim 10^{10}$ to 10^{11}sec^{-1}), the typical value of Q_e is approximately 10^9. Therefore, $n_{th} \approx 10^8 n_T^{-1}$. At helium temperature, $n_T \approx 10$ to 10^2, and consequently, $n_{th} \gg n_T$. This estimate shows that at $n \sim n_T$, such oscillators behave energetically in a quantum mechanical fashion. Of course, this statement does not contradict the fact that such oscillators obey classical laws in the mean (Dodonov, Manko, and Rudenko 1980). For the best mechanical oscillators with a frequency of $\omega \approx 10^4$ to 10^5 sec^{-1}, a quality factor larger than 10^9 is attained near liquid-helium temperatures (see chapter 2 for details). Therefore, n_{th} for these oscillators is also of the order $10^8 n_T^{-1}$. Since $n_T - kT/\hbar\omega \sim 10^6$ to 10^5 at $T - 4\text{K}$, then $n_{th} \approx 10^2$ to 10^3 in this case.

As this text is being prepared, nobody has yet succeeded in observing the quantum discreteness of energy in a high-Q macroscopic mechanical or electromagnetic system being driven by a heat bath. Nevertheless, experimental schemes have been proposed by Braginsky, Vorontsov, and Khalili (1977), Unruh (1978, 1979), and

Braginsky and Khalili (1980) to monitor the quantum features of an electromagnetic resonator at microwave and optical frequencies. These schemes rely on an interaction between the measuring device and the resonator that is proportional to the square of the resonator's amplitude. To achieve such a coupling, one can use ponderomotive effects at microwave frequencies, inverse Faraday effects, or optical detection effects at optical frequencies. For all these schemes, the back-action effects of the measuring procedure $(\Delta \mathcal{E} - \Delta n \, \hbar \omega)$ are significantly less than $\hbar \omega$, whereas the back-action effects on the oscillator's phase φ are substantial, in accord with the uncertainty relation $\Delta n \, \Delta \varphi \geq \frac{1}{2}$. The main obstacle to developing and using such procedures is the weakness of all known quadratically nonlinear couplings when the oscillation amplitudes are very small.

We turn now to quantum mechanical aspects of measurements of the position (or "generalized coordinate") of a high-Q oscillator. As is evident from equation (1.6), for a given averaging time $\hat{\tau}$ the random Brownian-motion changes of amplitude decrease as the dissipation is reduced (i.e., as the Q is increased). It is clear that there is a threshold value Q_{th} for the quality factor Q, at which for a given $\hat{\tau}$, T, and parameters of the oscillator, the thermal amplitude fluctuations become comparable to the coordinate width of a coherent-state wave packet. For a mechanical oscillator and an electrical circuit, this Q_{th} is given by

$$\Delta x_T - \left(\frac{kT\hat{\tau}}{M\omega Q_{\text{th}}} \right)^{\frac{1}{2}} - \Delta x_{\text{coh}} - \left(\frac{\hbar}{2M\omega} \right)^{\frac{1}{2}} \quad \text{mechanical} ,$$

$$\Delta U - \left(\frac{kT_e \omega_e \hat{\tau}}{C Q_{e\text{th}}} \right)^{\frac{1}{2}} - \Delta U_{\text{coh}} - \left(\frac{\hbar \omega_e}{2C} \right)^{\frac{1}{2}} \quad \text{electrical} . \qquad (2.5)$$

Here for the electrical circuit, C is the capacitance and ΔU is the voltage across the capacitor; ΔU plays the role of position (generalized coordinate). It follows from this formula that

$$Q_{\text{th}} \approx \frac{kT}{\hbar} \hat{\tau} - n_T \omega \hat{\tau} . \qquad (2.6)$$

Quality factors greater than Q_{th} have been achieved at temperatures $T \approx 4\text{K}$ and for reasonable averaging times $\hat{\tau}$ ($\sim 10^{-2}$ to 10^{-3} sec). It is clear that for position measurements with $Q \geq Q_{\text{th}}$, as for energy measurements with $n \leq n_{\text{th}}$, a quantum mechanical analysis is necessary. Such an analysis has been given by

Braginsky, Vorontzov, and Khalili (1978), Thorne *et al.* (1978), Holenhorst (1979), and others. The main conclusions of such an analysis are the following:

If an oscillator is coupled to a device that measures its position continuously and has a frequency bandwidth of order the oscillator frequency ω and an averaging time $\hat{\tau} \leq 2\pi/\omega$, then the minimum back-action perturbation of the measuring device on the position Δx_{BA} and the measurement error Δx_{meas} are approximately equal to Δx_{coh}

$$\Delta x_{BA} \approx \Delta x_{meas} \approx \Delta x_{coh} . \qquad (2.7)$$

From this it follows that the minimum impulsive force F_{min} (lasting a time $\hat{\tau} \leq 2\pi/\omega$) that can be detected by its influence on the oscillator is given by

$$F_{min}\hat{\tau} \approx 2\sqrt{\hbar M \omega} . \qquad (2.8)$$

This result is valid only if the position measurement is made continuously and over the full bandwidth associated with the short averaging time $\hat{\tau}$, in which case the result follows directly from Heisenberg's uncertainty principle for position and momentum. It is obvious that equations (2.7) and (2.8) hold only when the coupling between the oscillator and the measuring device is optimal. The accuracy of measurement will be worse in cases of nonoptimal coupling.

Thus, the classical formulae (1.5) and (1.7) for F_{min} are restricted by the quantum limit of equation (2.8) when the oscillator coordinate is detected continuously over a bandwidth of order ω and with an averaging time $\hat{\tau} \leq 2\pi/\omega$. If the continuous measurement of position is replaced by a suitably pulsed or modulated measurement (Braginsky, Vorontsov, and Khalili 1978; Thorne *et al.* 1978), the quantum limit F_{min} can be made much smaller than equation (2.8). (It can formally be made equal to zero if practical difficulties are not considered.) The appropriate procedures for measuring an oscillator's response to a weak force in this quantum regime will be analyzed in more detail in section 11, which is devoted to gravitational antennae. It is evident that formulae analogous to equations (2.7) and (2.8) can also be written for electromagnetic oscillators.

To summarize, quantum mechanical language is required to describe high-precision measurements of the energies or positions of oscillators with quality factors larger than some characteristic threshold value Q_{th}.

In conclusion, we will consider the question of whether the act of measuring an oscillator places limits on its quality factor. Suppose an observer were able to detect the damping of oscillations over a long period of time τ_Σ. Suppose also that the observer measured the oscillator's position at the beginning and end of the interval τ_Σ to an accuracy $\approx \Delta x_{\text{coh}}$, with the measuring device being out of contact with the oscillator during the entire time τ_Σ between the measurements. Since $\Delta x_{\text{BA}} \approx \Delta x_{\text{coh}}$ [equation (2.7)], the results of such a procedure enable the observer to determine the damping time τ^* if it is less than $\tau_\Sigma x_0 \sqrt{2M\omega/\hbar}$, where x_0 is the initial oscillation amplitude.

The expression for τ^* in terms of Q gives us the maximum measurable quality factor

$$Q_{\max} \approx \tau_\Sigma \left(\frac{\mathcal{E}\omega}{\hbar} \right)^{1/2} \approx \tau_\Sigma \xi \left(\frac{YSv}{\hbar} \right)^{1/2}, \qquad (2.9)$$

where $\mathcal{E}$ is the oscillation energy at the beginning of the interval τ_Σ. The last expression in this equation is specialized to the case of a low-order mechanical vibration of a solid cylinder or cube with amplitude of strain oscillation ξ, Young's modulus Y, cross-sectional area S, and speed of sound v. For $\tau_\Sigma - 10^7 \text{sec}$, $\xi - 10^{-3}$, $Y - 4 \times 10^{11} \text{Pa}$, $S - 10^{-2} \text{m}^2$, and $v - 10^4 \text{m sec}^{-1}$, equation (2.9) gives $Q_{\max} \approx 10^{28}$. As we shall see shortly, this estimate shows that from the point of view of the measuring procedure, there is nothing to prevent the detailed measurement of huge quality factors or of very small vibrational energy dissipation.

The analogous specialization of equation (2.9) to electromagnetic oscillations in a hollow cavity or in dielectric material gives the following limit for the maximum measurable quality factor

$$Q_{e\max} \approx \tau_\Sigma \left(\frac{\epsilon_0 E_0^2 Sc}{\hbar} \right)^{1/2}, \qquad (2.10)$$

where ϵ_0 is the dielectric constant, E_0 is the initial amplitude of the electric field oscillations, S is the cross-sectional area of the resonator, and c is the velocity of light. For $E_0 - 10^7 \text{volt m}^{-1}$, $S - 10^{-2} \text{m}^2$, and $\tau_\Sigma - 10^7 \text{sec}$, equation (2.10) gives $Q_{e\max} \approx 10^{29}$. As for mechanical resonators, this estimate of $Q_{e\max}$ is far larger than any quality factor ever obtained.

Quantum mechanical features of macroscopic high-Q oscillators play a significant part in two experimental programs: (1) the development of highly stable electromagnetic self-excited oscillators, and (2) the creation of highly sensitive gravitational-wave antennae. These experimental programs will be discussed in some detail in sections 9 and 11.

II   Mechanical Oscillators
with Small Dissipation

3. Fundamental Dissipative Process in Solids

Deformations that occur in the mechanical vibrations of a solid body are accompanied by processes that are not thermodynamically reversible. These processes are responsible for dissipation of the oscillation energy.

It is essential to discriminate between the energy dissipation that occurs in a perfect crystal lattice and that which occurs in a real, imperfect crystal or in an amorphous body. The dissipation in an ideal crystal can be regarded as "fundamental" and is nonremovable in principle. The dissipation in a real crystal or amorphous body is largely associated with lattice imperfections or with the internal structure of the body. When a real body vibrates, accompanying rearrangements of its defects and changes in their states cause a dissipation of mechanical energy. This chapter presents basic information about fundamental dissipative processes inherent in perfect crystals and about the most important processes peculiar to real crystals. Our discussion will not be comprehensive. (For further details and for discussions of dissipative processes not covered here see, for example, the monograph by Nowick and Berry 1972.) We begin our discussion in this section with dissipative processes that are fundamental.

Thermoelastic dissipation. Thermoelastic dissipation is caused by vibratory volume changes producing temperature changes that are spatially inhomogeneous. The resulting temperature gradients induce heat flow, which is accompanied by an entropy increase and a conversion of vibration energy into thermal energy. A calculation gives the following expression for the quality factor of a longitudinally oscillating bar with this type of loss (Landau and Lifshitz 1965)

$$\frac{1}{Q_{\text{Th.d.}}} = \frac{\kappa T \alpha^2 \rho \omega}{9C^2}, \tag{3.1}$$

where ω is the vibrational angular frequency, T is the bar's temperature, ρ is the density of the bar's material, C is the heat capacity per unit volume, α is the thermal expansion coefficient, and κ is the thermal conductivity of the bar's material. Equation (3.1) holds if the vibrations are adiabatic, which is the case for almost all frequencies of laboratory interest (up to 10 GHz). Though derived for an isotropic medium, equation (3.1) is correct in order of magnitude for an anisotropic crystal as well. Note that there is no thermoelastic dissipation in pure shear oscillations (e.g., torsional oscillations of a bar) because the volume does not change and hence there is no local oscillation of the temperature.

Dissipation due to phonon-phonon interactions. The anharmonicity of a crystal lattice leads to another fundamental dissipation mechanism, one caused by interactions between sound waves and thermal phonons. If the oscillatory (sound) wavelength is considerably larger than the mean free path of the phonons, we can regard the sound wave as locally changing the phonon frequencies and thereby perturbing the phonon distribution function away from its equilibrium Planck form. The process of restoring thermal equilibrium to the phonon gas is accompanied by dissipation of the sound-wave energy.

The theory of this process was first treated by Akhiezer (1938) and was developed further by Woodroof and Ehrenreich and by Bömmel and Dransfeld (1960). The first attempt to check the theory of phonon damping experimentally was made by Bömmel and Dransfeld using a quartz crystal. They found that the quality factor for a vibrating crystal is given by the expression

$$\frac{1}{Q_{\text{phph}}} = \frac{CT\hat{\gamma}^2}{\rho v^2} \frac{\omega \tau_{\text{ph}}^*}{1 + (\omega \tau_{\text{ph}}^*)^2}, \tag{3.2}$$

where v is the sound velocity, $\hat{\gamma}$ is Grüneisen's constant (which characterizes the anharmonicity of the crystal-lattice vibrations and is generally chosen to fit experimental data for the given temperature), and τ_{ph}^* is the phonon relaxation time, which can be identified with the thermal relaxation time in the expression for thermal conductivity

$$\kappa = \frac{1}{3} C v_D^2 \tau_{ph}^* . \tag{3.3}$$

Here v_D is the mean Debye sound velocity defined by the expression

$$\frac{3}{v_D^3} = \frac{1}{v_l^3} + \frac{2}{v_t^3} ,$$

v_l and v_t being the velocities of longitudinal and transverse sound waves in the solid body.

Other approaches have been proposed for evaluating the sound-wave attenuation in solid bodies caused by phonon-phonon interactions (see, for instance, Mason 1964 and Lemanov and Smolensky 1972). These approaches give expressions for the quality factor that are different from equation (3.2), but equation (3.2) is the simplest and most convenient for rough estimates.

Dissipation due to phonon-electron interactions in metals. In addition to the dissipation mechanisms already described, in metals there is a dissipation mechanism that depends on the presence of free electrons. A strain wave in a solid body sets up ion oscillations. Because of the positive ion charge, a varying electric field arises and forces the free electrons to move. The electron "gas" possesses a viscosity like a normal gas or fluid, in which there is a momentum exchange between layers moving at different velocities. For the model of free electrons with a spherical Fermi surface, and a metal with $\omega \bar{l}_e / v < 1$ ($\bar{l}_e$ is the mean free path of the electron), the resulting dissipation of longitudinal sound waves is given by the expression (Mason 1964)

$$\frac{1}{\omega Q_{ph\,e}} = \frac{8}{15} \frac{\mathcal{E}_F m_e \sigma}{\rho v^2 e^2} , \tag{3.4}$$

where m_e is the electron mass, e is the electron charge, σ is the electric conductivity of the metal, and $\mathcal{E}_F$ is the Fermi energy.

Table 1. The Value of ωQ for Fundamental Dissipative Processes in Solids

		Kind of dissipation process		
Material	T(K)	Thermoelasticity	Phonon-phonon interaction	Phonon-electron interaction
Quartz	300	6×10^{16}	10^{14}	—
(monocrystal)	4.2	1.5×10^{16}	4.5×10^{14}	—
Aluminum	300	6×10^{14}	3×10^{12}	5×10^{13}
	4.2	2×10^{16}	10^{14}	3×10^{11}

The fundamental dissipation mechanisms for solid bodies are those described above. These mechanisms can be regarded as fundamental because they occur even in perfect crystals. The quality factor estimates given by equations (3.1), (3.2), and (3.4) can be regarded as the limiting Qs for a resonator made from a given material.

It should be emphasized that for all of the dissipation mechanisms described above, the quantity Q^{-1} is proportional to ω if $\omega\tau_{ph}^{*} \ll 1$. Table 1 gives the values of ωQ determined by the fundamental dissipation mechanisms at two temperatures, 300 K and 4.2 K. Two typical materials are shown: quartz, which is widely used in high-Q mechanical resonators, and aluminum, which is also used in the construction of mechanical resonators. Table 2 gives the mechanical and thermal parameters for these materials, and also those for sapphire. The use of sapphire in the construction of high-Q resonators will be considered later in this chapter (Sec. 5).

It should be noted that various references give discrepant values of material parameters at low temperatures. The discrepancies are due in part to the fact that at low temperatures the physical properties of materials are strongly affected by defects and impurities. Therefore, the 4.2 K data in table 2 can be used for estimates only.

Figures 1 and 2 show experimental curves for the dependence of Q^{-1} on temperature for crystalline quartz resonators (Smagin 1974) and aluminum resonators (Suzuki, Tsubono, and Hirakawa 1978). These are the highest-Q results obtained to date. Using table 1 and introducing the eigenfrequencies of the resonators, one can compare these results with the limiting Q-values for these materials. The theoretical limits and the experimental data at low

Table 2. Mechanical and Thermal Parameters for High-Q Resonator Materials
(The first line gives the value at 300 K, the second at 4.2 K)

Parameter	Quartz (along z-axis)	Sapphire (along z-axis)	Aluminum (polycrystal)
ρ (kg m^{-3})	2.65×10^3 [1]	3.98×10^3 [3]	2.7×10^3 [7]
	—	—	—
v_l (m sec^{-1})	5×10^3 [1]	1.05×10^4 [3]	6.26×10^3 [7]
	—	—	—
C (Joule kg^{-1} K^{-1})	7.2×10^2 [1]	7.9×10^2 [4]	9×10^2 [7]
	1.9×10^{-1} *	1.2×10^{-5}	2.8×10^{-1}
κ (W m^{-1} K^{-1})	1.4×10^1 [2]	4×10^1 [5]	2.07×10^2 [7]
	3×10^1	3×10^2	1.3×10^3
α (K^{-1})	7×10^{-6} [1]	6.6×10^{-6} [6]	2.26×10^{-5} [7]
	2×10^{-8} *	4×10^{-11}	4×10^{-9} *
$\mathcal{E}_F$ (eV)	—	—	11.6 [8]
	—	—	—
σ (Ohm^{-1} m^{-1})	—	—	4×10^7 [7]
	—	—	6×10^9

*These values are obtained by extrapolation with the Debye law.
[1] Smagin and Yaroslavsky (1970). [2] Mason (1964). [3] Belyaev (1974). [4] Ginnigs and Furukawa (1953). [5] DeGoes and Dreyfus (1967). [6] Braginsky, Vasiliev, and Panov (1980). [7] Kikoin (1976). [8] Harrison (1970).

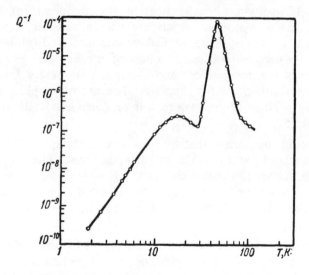

Fig. 1 Temperature dependence of Q^{-1} for quartz resonator with eigenfrequency of 1 MHz (Smagin 1974).

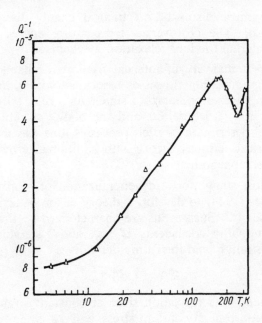

Fig. 2 Temperature dependence of Q^{-1} for aluminum resonator with eigenfrequency of 20 kHz (Suzuki, Tsubono, and Hirakawa 1978).

temperatures are in reasonable agreement. The product ωQ for aluminum at $4\,\mathrm{K}$ is approximately equal to $1.5 \times 10^{11}\mathrm{s}^{-1}$ (see figure 2) if $\omega = 2\pi \times 2 \times 10^{4}\mathrm{Hz}$. For lower mechanical frequencies the measured values of Q are higher but the product ωQ is lower. At $\omega = 2\pi \times 44\,\mathrm{Hz}$ and $T = 4\,\mathrm{K}$, Kimura *et al.* find $Q \simeq 4.0 \times 10^{7}$ corresponding to $\omega Q \simeq 1.1 \times 10^{10}\mathrm{s}^{-1}$. The experimental Q-value for quartz at $4.2\,\mathrm{K}$ is even better than the estimated theoretical limit. The explanation for this is that the value of $\omega\tau_{\mathrm{ph}}^{*}$ for the actual resonator is close to unity at this temperature, which violates the conditions for equation (3.2). The phonon attenuation at $\omega\tau_{\mathrm{ph}}^{*} \gg 1$ is described not by Akhiezer's theory, but by a theory developed by Landau and Rumer (1937) that we will not discuss here.

Dissipation due to lattice defects and the internal structures of solid bodies. Figures 1 and 2 show that at temperatures higher than liquid helium, the experimental Qs for quartz and aluminum resonators are much lower than the theoretical limits due to fundamental mechanisms of mechanical energy dissipation in a perfect crystal. The Q^{-1} versus temperature curves have peaks caused by "internal friction" relaxation mechanisms.

The great majority of internal friction mechanisms may be thought of as relaxation processes associated with time-dependent transitions to another equilibrium state. As a rule, they are associated with crystal lattice defects and the body's structure. For the mathematical description of such processes, one can use the equations of classical elasticity theory with additional terms that incorporate anelastic phenomena.

The most straightforward generalization of Hooke's law was used by Zener (1948) to develop a theory of the behavior of "standard linear solids." Such solids are characterized by three independent proportionality coefficients τ_ξ^*, τ_σ^*, and Y_R, which link the stress σ, the strain ξ, and their time derivatives

$$\sigma + \tau_\xi^* \dot\sigma = Y_R(\xi + \tau_\sigma^* \dot\xi) \,. \tag{3.5}$$

Here τ_ξ^* is the stress relaxation time at constant strain, τ_σ^* is the strain relaxation time at constant stress, and Y_R is the "relaxed elastic modulus" that governs the stress-strain relation after the completion of relaxation processes. The relation between quick changes $\Delta\sigma$ in stress and $\Delta\xi$ in strain is determined by the "unrelaxed elastic modulus"

$$Y_U = \frac{\Delta\sigma}{\Delta\xi} = Y_R \, \frac{\tau_\sigma^*}{\tau_\xi^*} \,.$$

In the case of a periodically oscillating strain, the relationship (3.5) between stress and strain produces a phase lag of the stress relative to the strain, and thereby the dissipation of mechanical energy. Putting periodically oscillating expressions for strain and stress in equation (3.5), we find for large values of the quality factor Q

$$\frac{1}{Q} = \frac{Y_U - Y_R}{\overline{Y}} \, \frac{\omega\overline{\tau}^*}{1 + (\omega\overline{\tau}^*)^2} \,, \tag{3.6}$$

where $\overline{Y}$ is the geometric mean of the two moduli Y_U and Y_R, and $\overline{\tau}^*$ is the geometric mean of the two relaxation times τ_ξ^* and τ_σ^*.

Thus, the energy losses in a vibrating "standard linear solid" are frequency dependent and are determined by equation (3.6). As a rule, the dissipation in solids is small, with a small difference between τ_ξ^* and τ_σ^*. Therefore, one can use a single relaxation time τ^*, not differentiating between τ_ξ^* and τ_σ^*. The quantity $(Y_U - Y_R)/\overline{Y}$ is known as the "intensity" of the relaxation process.

A plot of Q^{-1} versus $\log\omega\tau^*$ as given by equation (3.6) shows a peak, with a maximum at the point where $\log\omega\tau^* = 0$ (i.e., at $\omega\tau^* = 1$). This maximum is often called the "Debye peak" of losses.

This model of a standard linear solid makes an appropriate allowance for the behavior of real solids and has been applied successfully in descriptions and studies of a variety of relaxation processes in solids. In many relaxation processes a fundamental role is played by the movement of atoms from one location to another. In this case τ^* can often be described by the Arrhenius formula (Zener 1948)

$$(\tau^*)^{-1} = \tau_0^{-1}\exp\left(-\frac{\Delta\mathcal{E}_a}{kT}\right),\qquad(3.7)$$

where τ_0^{-1} is the rate factor, and $\Delta\mathcal{E}_a$ is the activation energy for the process. According to equation (3.7), τ^* varies greatly with temperature. Equations (3.6) and (3.7) explain the commonly observed features of vibration damping in solids. Peaks in Q^{-1} corresponding to $\omega\tau^* = 1$ arise at specific temperatures as the temperature is varied. By measuring the temperature of the peak at different frequencies, and by using equation (3.7), one can find the activation energy $\Delta\mathcal{E}_a$ of the process responsible for the dissipation. The knowledge of $\Delta\mathcal{E}_a$ helps one to identify the relaxation process.

Many relaxation phenomena in solids have been discovered and studied. As already mentioned, these phenomena are associated chiefly with defects in the crystal lattice. For instance, there is a specific type of defect that arises from the addition, removal, or substitution of one or several atoms. These are known as point defects. Lattice vacancies, impurities, and interstitial atoms are examples of this type of defect. When the crystal is deformed, point defects destroy the equilibrium of the distribution of atoms. The atoms and defects then try to move into a new equilibrium distribution, with a relaxation time τ^*. In a vibrating crystal this relaxation is accompanied by phase lags and dissipation. This relaxation process is known as strain reordering (Nowick and Berry 1972).

Dislocations are a more complex type of structural defect. Like point defects, dislocations can move through the crystal lattice. Deformations in a vibrating crystal cause the thermally activated motion of dislocations (dislocation relaxation). Various mechanisms dissipate the energy of moving dislocations; examples are thermoelastic effects and radiative damping. One of the features of a thermally activated relaxation process is its nonlinear character, which can result in an amplitude-dependent internal friction when the amplitude of deformation is large. The details of dislocation dynamics are far from fully understood.

Another important contributor to the damping of mechanical vibrations is stress relaxation along interfaces (the boundaries of units, grains, twins, and inclusions) (Nowick and Berry 1972). For instance, the elastic anisotropy and the chaotic orientation of grains in polycrystalline metals result in strain and stress fluctuations from one grain to another. Consequently, an applied oscillatory stress causes fluctuations of thermodynamic potentials in the grains; these fluctuations are attenuated by diffusion between grains. This also gives rise to various anelastic effects.

Let us consider in more detail another relaxation mechanism for mechanical energy dissipation: structural relaxation in fused quartz. Fused quartz is a remarkable material that is frequently used in the laboratory for constructing high-quality pendulums, tuning forks, and other mechanical oscillators. It is an easy material to work with. By its nature, quartz is dielectric, has a glass structure, and contains a disordered network of three-dimensional tetrahedrons. Because amorphous quartz has very low thermal conductivity ($\kappa = 1.2 \, \mathrm{W m^{-1} K^{-1}}$) and a small linear expansion coefficient ($\alpha = 0.5 \times 10^{-6} \, \mathrm{K^{-1}}$ at room temperature), it has a low level of thermoelastic losses. Mitrofanov and Frontov (1974) examined the attenuation of longitudinal sound oscillations in fused quartz at a frequency of 6.9 kHz. At room temperature, they obtained a quality factor of $Q = 4.4 \times 10^6$. However, studies at lower temperatures have revealed decreased Qs and a wide peak of attenuation in the 30 to 50 K region.

Figure 3 shows the low-temperature Q^{-1} diagrams obtained by Fine, Van Duyne, and Kenney (1954) for fused-quartz resonators. The peak of attenuation is of the relaxation type, but its width is much larger than in the case of a simple Debye peak. This indicates that the relaxation process has a wide spectrum of activation energies. An explanation of the attenuation mechanism

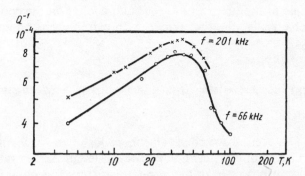

Fig. 3 Temperature dependence of Q^{-1} for fused quartz resonators with eigenfrequencies of 66 and 201 kHz (Fine, Van Duyne, and Kenney 1954).

offered by Anderson and Bömmel (1955) is connected with the structure of fused quartz. The bonds Si–O–Si in the disordered network of three-dimensional tetrahedrons are slightly bent, since the oxygen atom possesses two equilibrium positions separated by a small potential barrier. Thermal activation by sound waves causes transitions of oxygen atoms between these two equilibrium positions, giving rise to dissipation of the wave energy. The spectrum of activation energies is determined by the distribution of bond angles. This mechanism of energy loss is created by the structure of the fused quartz and consequently cannot be removed. Thus, quartz cannot be used when it is necessary to have low acoustic losses at low temperatures.

The most important conclusions of our brief review of processes that dissipate mechanical energy in solids are the following. The experimentally observed attenuation in solids is usually produced by relaxation processes associated with defects of structure. Nevertheless, there are a number of materials in which the energy dissipation at low temperature is almost solely determined by "fundamental" dissipation processes (i.e., by processes intrinsic to perfect crystals). Dielectric monocrystals with small numbers of defects and low mobility of defects fall into this category. A low mobility of defects is also characteristic of extremely brittle materials, such as TiC, SiO_2, Si, TiO_2, and Al_2O_3 (Libowitz 1976). Reducing the temperature to 4.2 K and below not only lowers the mobility of defects, but also decreases the fundamental dissipation due to phonon-phonon interactions. In section 5 we shall discuss an example of the successful use of such cooling in the construction of a mechanically vibrating system with extremely small energy dissipation.

4. Energy Losses Arising from the Design of a Mechanical Resonator

In this section we consider energy losses in mechanical resonators that are due to technological and design factors rather than to dissipation processes in the resonator material. Among these are losses caused by transmission of power to the surrounding gas, losses in suspension devices, losses in surface layers of the resonator, and losses caused by energy exchange between normal modes of vibration. Eliminating or diminishing these losses is the main problem in designing high-quality mechanical resonators made from materials with small internal losses. The extent of the energy losses in a resonator is a function not only of its eigenfrequency and temperature but also of the details of its design. We will consider an oscillator of the simplest kind, a cylindrical bar that is excited in its fundamental mode of longitudinal vibration.

Losses due to gas friction. A mechanical oscillator vibrating in a gaseous medium creates sound waves that carry its power away. An evaluation of this process gives the following expression for the Q (see, for instance, Gorelik 1959)

$$\frac{1}{Q_{sw}} = \frac{2P}{\pi\rho v}\left[\frac{C_P}{C_V}\frac{\mu}{RT}\right]^{1/2}, \tag{4.1}$$

where R is the universal gas constant, ρ is the density of the bar's material, v is the velocity of sound in the bar, μ is the mean molecular weight of the gas, and P is the gas pressure.

Equation (4.1) holds when the mean free path of the gas molecules is considerably less than the wavelength of the sound waves in the gas. For instance, this formula is appropriate for oscillator frequencies of 10 kHz at room temperature and at gas pressures $P > 10^{-3}$ Torr. At gas pressures lower than this, one cannot compute the dissipation using a model in which the mechanical vibrations produce sound waves in the gas. Instead, one can estimate the losses by regarding the gas as an ensemble of ideal, noninteracting molecules. When it collides with the surface of the vibrating resonator, each gas molecule acquires an additional momentum mv_{surf} due to the resonator's motion, where m is the mass of the molecule, and v_{surf} is the vibrational velocity of the resonator's surface at the collision point. Given the number of molecules colliding with a surface element ΔS per unit time, one can estimate the momentum lost by the surface element and the resulting force on it:

$$F = \frac{1}{4} n \bar{v} m \, v_{surf} \Delta S \, , \tag{4.2}$$

where $\bar{v}$ is the mean speed of the thermal molecular motion and n is the number of molecules per unit volume.

For a cylindrical resonator with length L and diameter D, equation (4.2) produces the following estimate for the quality factor in a rarefied gas

$$\frac{1}{Q_{gas}} = \frac{P}{\pi \rho \bar{v}} \left(1 + \frac{L}{D} \right) \left(\frac{8\mu}{\pi RT} \right)^{\frac{1}{2}} . \tag{4.3}$$

For instance, for a quartz resonator which has a length equal to its diameter and is suspended in a vacuum chamber with a residual gas pressure of $\sim 10^{-4}$ Torr at room temperature, equation (4.3) yields the estimate $Q_{gas}^{-1} = 3 \times 10^{-12}$. It obviously follows that gas-friction losses become very small even in a modest vacuum.

Surface losses. Losses in the surface of a mechanical oscillator can be substantial. The properties of the surface layers differ sharply from those of other parts of the crystal. Defects — irregularities, microcracks, spallings — are caused by machining (cutting, grinding, polishing). Sometimes the surface is modified by so-called "surface cold working." This produces two layers of crystal surface. The top layer is several microns thick and polycrystalline in nature. It is composed of a large number of randomly oriented, extremely small discrete crystallites. It is also saturated with atoms of carbon that diffuse through it when the surface is polished with diamond paste. Below this is a second layer, hundreds of microns thick, with an elevated density of dislocations. The layer thicknesses indicated here are typical, but the exact thicknesses depend on the properties of a given crystal and on the machining procedure used (see, for example, Achmatov 1963). Under ordinary conditions there are layers of absorbed molecules of gases, water, and organic compounds in addition to these two layers.

Of course, mechanical deformations of a system as complex as a surface are accompanied by many relaxation processes. In the case of a vibrating resonator, deformations upset the thermodynamic balance in the surface, and the restoration of this balance is accompanied by irreversible processes that give rise to the dissipation of elastic energy. These processes can be regarded as a "viscosity" of the surface layer.

The creation of heat flow in the surface layer's chaotically distributed crystallites is one mechanism of energy loss. Dissipation due to this process was first estimated by Zener (1948); Smagin and Yaroslavsky (1970) regard this process as the primary loss mechanism in the surface layers of quartz resonators.

If the length λ of a sound wave traveling in a resonator is much larger than the size a of a crystallite, one can regard the resonator material as subjected to a temporally oscillating but spatially homogeneous pressure. However, because of the random shapes and chaotic orientations of the crystallites, the boundary conditions on a crystallite are not uniform. This produces inhomogeneous heating and temperature gradients inside each crystallite. The temperature then gets homogenized by heat conduction in a time $\tau_T^* \sim a^2/D_T$ (thermal equilibrium relaxation time), where the thermal diffusion coefficient D_T is equal to κ/C_p. The "intensity" of this thermal relaxation process is

$$\frac{Y_U - Y_R}{Y} \sim YT \frac{\alpha^2}{C_P}, \tag{4.4}$$

where Y is Young's modulus for the crystallite material.

Consider a cylindrical resonator of length L and diameter D, vibrating in a longitudinal mode. Losses in the surface layers, which are composed of chaotically oriented crystallites, are given by the standard relaxation equation (3.6). We assume that losses in the interior are negligible and that the only significant losses are in the side surfaces. The resonator's quality factor is then determined by the ratio of the power dissipated in the side surface layers to the longitudinal vibration energy. Assuming that the deformation is uniform along the cross sections of the resonator, and using equations (4.4) and (3.6), one gets the following expression for the Q of a resonator with a damaged layer of depth h

$$\frac{1}{Q_{\text{surf}}} \approx \frac{4h}{D} \frac{YT\alpha^2}{C_P} \frac{\omega\tau_T^*}{1 + (\omega\tau_T^*)^2}. \tag{4.5}$$

Here the factor $4h/D$ is the ratio between the volumes of the damaged layer and the resonator as a whole, and $\tau_T^* = a^2 C_P/\kappa$ depends on the size of the crystallites.

A distinctive characteristic of any relaxation process is its peak of vibration damping at $\omega \tau_T^* = 1$. This peak is smeared out in the case of surface losses due to the large range of crystallite sizes. One can make a rough estimate of the surface losses for a sapphire resonator using equation (4.5). For crystallite size $a \approx 10^{-4}$ cm, depth of damaged layer $h \approx 10^{-2}$ cm, resonator diameter $D \approx 4$ cm, resonator frequency $\omega/2\pi \approx 30\,\text{kHz}$, and room temperature, one obtains $Q_{\text{surf}}^{-1} \approx 1.5 \times 10^{-7}$ for the contribution of surface losses to the Q. The measured value of Q^{-1} for a monocrystal sapphire resonator in our laboratory with a resonant frequency of 34 kHz was 2×10^{-7} after polishing with boron carbide having an abrasive ~ 50 μm grain. The value of Q^{-1} dropped to 10^{-8} after polishing with a diamond paste with grains 4 to 5 μm in size. Note that surface losses make a substantial contribution to vibration damping, not only in a monocrystal resonator but also in resonators made from other materials, in particular, fused quartz (Mitrofanov and Frontov 1974). As is evident from equation (4.5), surface losses associated with thermal currents in crystallites decrease with reduced temperature and become very small at liquid-helium temperatures.

Films absorbed on resonator surfaces create additional energy losses, and therefore stringent standards are set for the cleanliness of high-Q resonators.

Effective physicochemical techniques have recently been developed for the surface treatment of various materials. With these techniques, damage to the crystal lattice near the crystal surface can be minimized. The strength of a specimen can be regarded as an indirect indicator of the quality of its surface treatment. It is known, for instance, that the strength of a specimen under tension is significantly dependent on the condition of the specimen's surface (i.e., damage to its surface layer). Table 3 (Libowitz 1976) gives the maximum breaking strength for sapphire rods at room temperature after surface treatment by various techniques.

Losses due to coupling of different kinds of vibrations. Designing a mechanical resonator usually entails an evaluation of its various eigenfrequencies of oscillation. For resonators with

Table 3. Maximum Breaking Strength of Sapphire Bars at Room Temperature
After Various Treatments

Treatment technique	Strength (10^9 Pascal)
Flame polishing, selected working area of specimen (1600°C)	7.35
Boron etching	6.86
Machine polishing, firing in oxygen (1600°C)	1.04
Annealing, machine polishing	0.78
Lapping	0.59
As manufactured (no additional treatment)	0.44

simple shapes (rods, plates, etc.), the vibration frequencies of the
fundamental modes (longitudinal, torsional, and shear) are given
with sufficient accuracy by well-known classical formulas. For
instance, for a rod whose length is much greater than its diameter,
the frequencies ω_n of low-order longitudinal modes can be
evaluated in a one-dimensional approach allowing for only longitudinal strains; the resulting formula is well known:

$$\omega_n = \frac{n\pi}{L} v ,$$

where L is the length of resonator and $v = (Y/\rho)^{1/2}$ is the longitudinal sound velocity.

In order to estimate losses in the mounting of a resonator, the
stress-strain distribution in the mounting must be known. Finiteness of the diameter of a cylindrical rod leads to a dependence of
the longitudinal strain on distance from the rod's axis. If the rod
material is isotropic, the amplitude of longitudinal displacement
achieves its maximum value x_{oz} in the centers of its end faces,
whereas the maximum value of the amplitude of radial displacement x_{or} occurs on the outer circumference of its middle cross section. The ratio of these two maximum displacements is approximately (Rayleigh 1929)

$$\frac{x_{oz}}{x_{or}} = \frac{\pi v_0 D}{2L} , \qquad (4.6)$$

where v_0 is the Poisson ratio of the rod material.

If the rod is made of a monocrystal, crystal anisotropy produces a much more complicated picture. The relation between the

strain tensor ξ_{kl} and the stress tensor σ_{ij} in an anisotropic medium is determined by an elasticity tensor c_{ijkl}

$$\sigma_{ij} = c_{ijkl}\,\xi_{kl} \ .$$

For a crystal, symmetries reduce the number of independent components c_{ijkl}. In particular, there are three independent elasticity components for silicon and six for quartz and sapphire. Because of the combined influence of the various elasticity components, waves propagating in an infinite crystal cannot be purely transverse or purely longitudinal, except for very special propagation directions. In other words, the angle between the direction of the displacement vector and the normal to the wave front cannot be equal to zero or $\pi/2$. Similarly, in a cylindrical rod vibrating in its fundamental longitudinal mode, the nodal surface for longitudinal motions does not intersect the rod axis at right angles unless the axis is oriented along the direction of maximal or minimal Young's modulus (Cady 1946).

For these reasons it turns out that a crystalline rod, in which longitudinal vibrations are excited, has no points at which all components of the displacement vector vanish. This complicates the design of suspension systems that minimize losses.

Because of the connection between deformations in different directions in a crystal (due to the anisotropy of the elasticity c_{ijkl}) there is a coupling between vibrations of different simple types; i.e., between what one might otherwise have thought were distinctly different normal modes. The strength of this mode-mode coupling is determined both by c_{ijkl}-induced coupling coefficients and by the frequency differences of the simple modes. Strong coupling arises if the modes have nearly equal frequencies. In this case, significant energy transfer from one simple mode to another is possible, and the quality factor of a chosen simple mode drops accordingly.

This effect has been observed experimentally. A cylindrical sapphire resonator 45 mm in diameter and 152 mm long was tested in the authors' laboratory. The fundamental-mode frequency was $f = 34014\,\mathrm{Hz}$, and the quality factor was measured to be $Q = 1.2 \times 10^8$. After this measurement, facets 18 mm wide were cut out on the ends of the crystal, making angles of 6° to the central axis. The resonator was then tested again and found to have two simple modes of nearly equal frequency ($f_1 = 34260\,\mathrm{Hz}$ and $f_2 = 35068\,\mathrm{Hz}$). One of these modes entailed vibrations along the axis and the other vibrations along diagonals. The quality

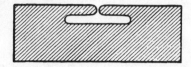

Fig. 4 Sapphire gravitational-wave antenna (seen in profile).

factor for each of these modes was measured and found to be $Q \approx 4 \times 10^7$. Thus the existence of a mode with a frequency close to that of the resonator reduced the Q in this experiment by approximately a factor of three.

Of course, not every deviation from a regular form (cylindrical, for instance) gives rise to damping. Figure 4 shows a variant of a high-Q gravitational wave antenna constructed in the authors' laboratory from a sapphire cylinder with two cut-out "horns." This antenna profile was designed and constructed so that the resonant frequencies of the horns would be maximally removed from the fundamental resonant frequency of the cylinder. All changes in cross section were made as smooth as possible. Before the horns were cut out, the Q of the fundamental mode was 1.8×10^8 at 78 K; afterwards it was 1.2×10^8. The small reduction in Q was probably due entirely to inadequate polishing of the horns.

Losses in a resonator's suspension. Keeping suspension losses small is an especially complicated problem in the design of high-Q mechanical resonators. Useful insight has come from the construction of quartz resonators and experiments with them by Smagin and Yaroslavsky (1970). The smallest suspension losses were obtained with lens-shaped quartz resonators vibrating in shear modes.

As mentioned earlier, a cylindrical resonator vibrating in the fundamental mode has no point at which all components of displacement vanish, but the displacement is minimal at the central cross section. Thus, it is natural to support the resonator there. In experiments in our laboratory the most successful suspensions were made with fiber loops of thin wire or silk thread (figure 5a). Such suspensions can be stable if the resonator is balanced accurately, if the loop plane passes through the center of gravity of the resonator, and if the points of tangency where the fiber contacts the resonator

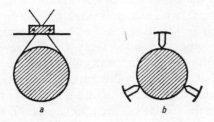

Fig. 5 Two types of suspension for cylindrical resonators.

are above the center of gravity.

Another type of suspension is a rigid mounting of the type shown schematically in figure 5*b*. The resonator is held in its central plane on three centers placed at 120° angles to each other. A comparison between suspension on centers and suspension with a loop of tungsten wire was made for a ruby resonator in our laboratory. With suspension on centers, the quality factor was 10^6, while for suspension with a tungsten wire it was 11×10^6. When rubber gaskets were placed between the suspension centers and the resonator, the quality factor increased but only to 4×10^6. The maximum quality factor that has been achieved for a sapphire resonator on a fiber-loop suspension is 4×10^8 at room temperature, and 5×10^9 at liquid-helium temperature.

The first investigations and estimates of energy losses in fiber-loop suspensions were made by Bordoni (1947), who used a simple model. Because of the finite value of the Poisson ratio, longitudinal vibrations of a cylindrical resonator produce transverse oscillations in the nodal plane and a resulting motion of the suspension fiber. This motion excites an elastic wave in the fiber that carries away part of the resonator's vibration energy. According to this model, the energy flux in the fiber's elastic wave determines the resonator's suspension losses.

The effects of elastic-wave reflections must be taken into account when estimating the energy flux along the fiber and into surrounding bodies. For example, to estimate the suspension losses for a longitudinally vibrating mechanical resonator supported as in figure 5*a*, we assume that the amplitudes of the fiber vibrations at the points of contact with the resonator are equal to the amplitudes of displacement of the corresponding points of the resonator. We

assume also that the other end of the fiber is attached to an absolutely rigid support, which means that the elastic-wave energy is
entirely dissipated in the fiber itself. This is a simplifying assumption, since the support is never truly rigid. Nevertheless, it is a
good approximation if the mechanical impedance of the support
$Z_{sup} \approx j\omega M_{sup}$ (where M_{sup} is the support's mass) is much greater
than the mechanical impedance of the fiber. The maximum possi-
ble value for the fiber's impedance is $Z \approx Z_0 Q_f$, where
$Z_0 \approx \rho_f v_f S_f$ is the wave resistance of a fiber that has density ρ_f,
sound velocity v_f, cross sectional area S_f, and quality factor Q_f.
For example, a support with mass 1 kg can be approximated as
absolutely rigid for a resonator with frequency $\sim$10 kHz supported
by a fiber with diameter $\sim$10 μm and quality factor $Q_f \sim 10^3$.
Under these conditions the vibrations of the support are negligible.

A straightforward calculation using this model gives the following formula for the influence of suspension losses on the
resonator's quality factor

$$\frac{1}{Q_{susp}} = \frac{2\rho_f v_f S_f x_{ext}^2}{M_{eq}\omega x_0^2} \frac{\omega l}{v_f Q_f}\left[1 + \frac{1}{2}\left(\frac{\omega l}{v_f Q_f}\right)^2 - \cos\frac{2\omega l}{v_f}\right]^{-1}. \quad (4.7)$$

Here M_{eq} is the equivalent mass of the resonator, l is the length of
the fiber from its point of first contact with the resonator to the
point where it is attached to the support, x_0 is the amplitude of the
resonator's vibrations, and x_{ext} is the amplitude of displacement of
the resonator surface at the point of first contact with the fiber. If
the resonator is suspended in its central plane, then x_0 is related to
x_{ext} as x_{oz} is related to x_{or} in equation (4.6). If the resonator is
suspended by two loops near its end faces, then $x_0 = x_{ext}$.

Equation (4.7) exhibits resonance effects in the suspension
fiber. Periodic variations of the resonator Q due to varying fiber
lengths have been observed experimentally in our laboratory.
However, there is only a qualitative agreement between equation
(4.7) and the experimental data for sapphire resonators. The measured Qs are larger than the calculated ones by almost a full order
of magnitude. Evidently the assumptions underlying the derivation
of equation (4.7) are not quite correct. In particular, resonator
vibrations at the point of contact with the suspension fiber are not
completely transmitted to the fiber. The transmission of the resonator energy to the fiber turns out to depend strongly on the state
of the surfaces of the fiber and of the resonator. Suspension with a
polished wire produces damping almost two times weaker than that

achieved with an unpolished wire. The presence of a fatty film (e.g., pork fat) at the points of contact between the suspension fiber and the resonator is important. This film acts as a lubricant, diminishing the coupling between the fiber and the resonator. Usually the fatty film is specially applied before suspending the resonator. If the points of contact were thoroughly cleaned, there would be a sharp drop in the quality factor.

Two other aspects of the coupling between the resonator and the suspension fiber are worth noting. First, the coupling decreases with decreasing temperature, and hence the suspension losses are also reduced. Second, if the resonator is heated after being cooled to liquid-helium temperature and is then cooled again with no change of suspension, its quality factor usually increases by about 20%. These phenomena demonstrate the complexity of the processes that occur at the points of contact. Thus far it has been impossible to describe these processes by a simple model.

It is interesting also to note that in our experiments with a sapphire resonator at 4.3 K, although the Q was 5×10^9 if the resonator was suspended at its central plane by one loop of silk thread, suspension near the center by two silk-thread loops 8 mm apart gave a Q of only 1.2×10^9.

In this section we have mainly discussed fiber-loop suspensions. This kind of suspension is convenient only if there is no need to fix the resonator rigidly.

5. High-Q Resonators Made from Sapphire Monocrystals

Figure 6 shows the highest quality factors obtained thus far for audio-frequency and ultrasonic-frequency resonators made from the aluminum alloy Al-5056, niobium, silicon, and sapphire. Most experimental Qs that are close to the limits fixed by fundamental loss mechanisms are achieved with mechanical resonators made from monocrystals with small dislocation mobilities, for example, silicon, quartz, and sapphire. As figure 6 shows, sapphire is the best material of all. The theoretical analysis in section 3 indicates the reasons for expecting sapphire to be an excellent material from which to construct high-Q resonators. It should be noted that large, high-grade sapphire monocrystals are available from industrial sources. Extensive applications for sapphire have been found in experimental physics and engineering (Belyaev 1974).

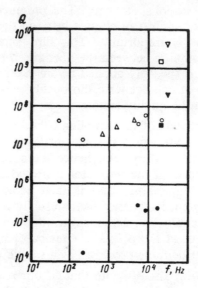

Fig. 6 Maximum quality factors for mechanical resonators made from different materials. O = Al-5056 (Oide, Tsubono, and Hirakawa 1980), △ = Nb (Blair *et al.* 1980b), □ = Si (McGuigan *et al.* 1978), ▽ = Al$_2$O$_3$ (Bagdasarov *et al.* 1977). Solid symbols correspond to $T = 300$ K, open symbols to $T = 4.2$ K. Note added in press: Recently Veitch *et al.* (1985) have reported a Q of 2×10^8 at 4 K in a niobium bar.

This section is devoted primarily to methods for constructing high-Q sapphire resonators. The damping of elastic vibrations in sapphire is considered in detail and the results of experimental investigations of sapphire resonators are reported.

Sapphire is the crystalline form of corundum $\alpha-Al_2O_3$ belonging to the trigonal system ($\bar{3}m$ symmetry class). The corundum crystal lattice consists of bivalent ions of oxygen and trivalent ions of aluminum. The two varieties of corundum, sapphire and ruby, derive their different colors from either isomorphous impurities of Cr_2O_3 (red ruby), or colloidal impurities of certain metals (blue sapphire).

By the term "sapphire" we mean leucosapphire (i.e., corundum with minimum impurities of heavy metals). Sapphire is a hard material with a high melting point (its Moh scratch number is 9; its melting point is 2030°C). It has a high chemical resistance but can be etched at high temperatures with a restricted set of chemical agents (Belyaev 1974). These properties are conducive to the fabrication of sapphire resonators with highly stable characteristics. Sapphire has one of the highest Debye temperatures ($T_D = 1047$ K) and is easily machined with diamond tools.

Special electromechanical transducers are necessary to excite vibrations in sapphire because sapphire has no piezoelectric or piezomagnetic properties. One of the main requirements here is that the damping induced by the transducers must be much less than the natural damping of vibrations in the resonator. Noncontacting capacitance transducers satisfy this requirement.

Figure 7 is a schematic drawing of a longitudinal vibration resonator with driving and readout transducers. The cylindrical or block-shaped resonator is supported at its central plane, which contains its center of gravity. The end faces of the resonator are covered with thin (about $1\,\mu$m) metal layers by means of evaporation or chemical deposition. A metal plate is placed near each end of the resonator, separated from the metal layer by a gap of about 0.5 mm. Together with the metal layers, they form a pair of capacitors, one at each end of the resonator. When an alternating voltage $U = U_0 \sin \omega t$ is applied to the driving pair of plates, an electric force between the plates and the metal layer excites elastic vibrations in the resonator. The driving force on the resonator's face is

$$ F = \frac{\epsilon_0 S U^2}{8 d^2}, \tag{5.1} $$

where ϵ_0 is the dielectric constant, S is the area of the metal layer, and d is the distance between the plates and the resonator. The difference between the numerical coefficient in this formula and that in the well-known formula for the electrostatic force between two capacitor plates is caused by the three-plate structure of this capacitor. The electric force [equation (5.1)] is proportional to the square of the voltage applied to the plates. If the voltage varies sinusoidally, it must be accompanied by a constant voltage U_c in order to produce a component of the force at the frequency of the applied voltage.

The relaxation time τ^* of high-Q sapphire resonators can be several hours long. This affects the choice of the procedure used to initiate vibrations in the resonator. A transient excitation of vibrations is needed, for instance, in order to measure the resonator Q by observing the damping of free vibrations. If one wants to

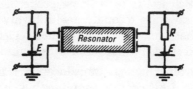

Fig. 7 A mechanical resonator with capacitive transducers.

produce a steady vibration amplitude by applying a resonant force, one should keep the frequency of the driving voltage and the frequency of the mechanical oscillator equal during a time $\hat{\tau} > \tau^*$. It is difficult to achieve this with commonly used electrical oscillators, owing to their inadequate frequency stability and frequency control. For this reason the excitation of vibrations is normally performed in a nonstationary way.

Suppose that the frequency of the exciting voltage is exactly equal to the frequency ω_0 of some vibration mode of the resonator. Let the voltage be applied for a time $\hat{\tau} \ll \tau^*$. In the case of the fundamental mode, the resonator can be regarded as a lumped oscillator with equivalent mass $M_{eq} = M/2$ and spring constant $K_{eq} = M\omega^2/2$. The vibration amplitude attained during the time $\hat{\tau}$ is then

$$x_0 = \frac{\epsilon_0 S U_0 U_c \hat{\tau}}{4 d^2 M \omega_0}. \tag{5.2}$$

A conversion of the resonator's mechanical vibrations into electrical signals is performed by the transducer that is placed at the other end of the resonator and that functions as a capacitance microphone. The vibrations of the resonator change the distance between the end of the resonator and the plates fixed beside it, causing the capacitance of the plates plus resonator end to vary sinusoidally.

In constructing sapphire resonators, it is possible to use capacitive transducers of this type without metal layers on the end faces because of the high value of the dielectric constant of sapphire ($\epsilon_{\parallel} = 10.5$; $\epsilon_{\perp} = 8.6$). In a nonuniform electric field (which arises when voltage is applied to the transducer plates), the naked resonator experiences a force that can excite vibrations. Conversely, the displacement of the dielectric near the plates changes the capacitance between the plates. However, without the conducting layer on the end face of the resonator, the strength of the coupling is reduced. These simple capacitive transducers can excite and display sapphire resonator vibrations with amplitudes as small as 10^{-6} to 10^{-9} cm.

It is clear that capacitive transducers induce additional rigidity (spring constant) and damping in the resonator. These effects can be estimated easily

$$K_{ind} = -\frac{\epsilon_0 S U_c^2}{4d^3} \frac{1}{1 + R^2 C^2 \omega^2}, \tag{5.3}$$

$$\frac{1}{Q_{ind}} = \frac{\epsilon_0 S U_c^2}{2d^3 M \omega^2} \frac{RC\omega}{1 + R^2 C^2 \omega^2}, \tag{5.4}$$

where U_c is the constant voltage applied to the transducer, C is the average capacitance of the transducer, R is the resistance in the transducer circuit, S is the area of the metal layers on the ends of the resonator, d is the distance between the transducer plates and the resonator, and M is the mass of the resonator.

Let us estimate the influence of losses in a transducer on the quality factor of a sapphire resonator. To get maximal coupling, the parameters of the transducer are chosen so that the inequality $RC\omega \gg 1$ is satisfied. For a resonator and transducer with $M = 1$ kg, $f = 3 \times 10^4$ Hz, $d = 5 \times 10^{-4}$ m, $S = 10^{-3}$ m^2, $R = 50$ MOhm, and $U_c = 100$ V, equation (5.4) gives $Q_{ind}^{-1} \approx 10^{-12}$.

The metal layers on the ends of the resonator (with a thickness no greater than $1\,\mu$m) cause an additional loss in the resonator of magnitude $Q^{-1} \lesssim 10^{-11}$. This has been confirmed experimentally.

We have considered above only the simplest scheme for converting mechanical resonator vibrations into electrical signals. When using a resonator in a physical experiment where extremely small influences on the resonator ($\delta x_0 \ll 10^{-9}$ cm) must be monitored, one can instead employ highly sensitive parametric type transducers (Braginsky and Manukin 1974).

We will now describe in detail the results of experimental studies of quality factors for sapphire resonators performed in our laboratory at Moscow State University. We describe first a series of experimental runs that showed how the value of Q^{-1} depends on temperature (Bagdasarov et al. 1974). The resonator was made

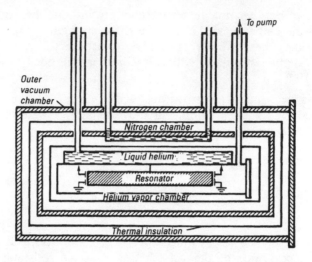

Fig. 8 A cryostat for low-temperature experiments with mechanical resonators.

from a sapphire monocrystal in the form of a cylinder with a length of 137 mm and a diameter of 44 mm. The axis of the cylinder was placed at an angle of 60° to the optic axis of the crystal. The frequency of the fundamental mode was equal to 38 kHz.

The crystal surface was abraded with a plastic that was covered with successive layers of diamond pastes having gradually finer grains. The rubbing was completed with a 7 to 10 μm grain paste. The crystal was then polished using felt with a 3 to 5 μm grain diamond paste. After abrading and polishing, the crystal was placed in a molybdenum container and annealed in vacuum at a temperature of 1100°C. Then the crystal was exposed to a boiling solution of $5HNO_3 + 3H_2SO_4 + 2H_2O$ for an hour to remove the molybdenum deposit from the crystal surface. This treatment helped to diminish losses in the surface layers of the crystal.

The crystal was suspended with silk thread (or, alternatively, with 100 μm tungsten wire) on a special support (see Section 4). The support with the suspended resonator was placed in a cryogenic vacuum chamber, the arrangement of which is shown in figure 8. The resonator was cooled to a temperature of 4.3 K by exchanging heat with gaseous helium at a pressure of ~0.1 Torr. Then the exchange gas was pumped out, and measurements were

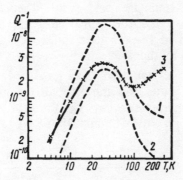

Fig. 9 Calculated (curves 1 and 2) and experimental (curve 3) temperature dependences of Q^{-1} for a sapphire resonator with vibration frequency of 38 kHz.

taken at a pressure of less than 10^{-5} Torr.

The quality factor was determined by measuring the damping time for free vibrations. Capacitive transducers were used to excite and monitor the vibrations. The temperature of the resonator was determined from the observed vibration frequency, using a temperature/frequency relation measured in advance with a "copper-constantan" thermocouple and a carbon-base resistor thermometer. The thermocouple and thermometer contacted the crystal during the calibration measurements, but there was no such contact during the measurements of damping time.

Figure 9 shows Q^{-1} for the longitudinal vibrations as a function of the temperature of the sapphire resonator. There is an attenuation peak at about 30 K, the same temperature as the well-known peak in the thermal conductivity of sapphire (De Goes and Dreyfus 1967). Since according to Akhiezer's theory the attenuation of sound waves is proportional to the relaxation time of thermal phonons and thereby to the thermal conductivity, the observed attenuation peak can be understood as the result of Akhiezer's process of phonon-phonon coupling.

The damping of longitudinal vibrations in sapphire as a function of temperature can be estimated with the help of equation (3.2) with $\hat{\gamma}^2 = 2$ (the average value of Grünisen's constants for sapphire as obtained from thermal-measurement data). The calculated damping is given in figure 9 by the dashed curve labeled 1.

The experimental data lie below this theoretical curve. The discrepancy between the theoretical and experimental results is large, and experiment reveals that the actual damping is weaker than that predicted by theory.

As noted above, equation (3.2) gives only a rough estimate. A more accurate value for the sapphire quality factor due to phonon-phonon interactions can be found from the expression (Mitrofanov, Ovodova, and Shiyan 1980)

$$\frac{1}{Q_{ph\,ph}} = \frac{\sum_i C_i T[\gamma_i^2 - \overline{\gamma_i^2}]}{\rho v^2} \frac{\omega \tau_{ph}^*}{1 + (\omega \tau_{ph}^*)^2}. \tag{5.5}$$

Here ρ is the density of the resonator material, v is its sound velocity, τ_{ph}^* is the mean relaxation time for thermal phonons as determined by equation (3.3), C is the heat capacity per unit volume of the i'th phonon mode in the crystal, γ_i are Grünisen's coefficients for the mode's strain, and $\overline{\gamma_i}$ are quantities obtained by averaging γ_i over all directions in the crystal. For longitudinal vibrations along the x-axis of the crystal

$$\gamma_i = -\frac{\partial v_i / \partial \xi_x}{v_{i0}},$$

where v_{i0} are the frequencies of phonon modes in the nonstrained state of the crystal. Grünisen's coefficients can be calculated for various directions in the crystal if the third-order elastic moduli are known (Brugger 1964). The sum in equation (5.5) can be evaluated approximately by computing Grünisen's coefficients for some set of directions in the crystal and performing the sum under the assumption that the values of C_i are direction independent.

For deformation along the x-axis of the crystal, the calculated value of Grünisen's coefficient is $\gamma_i = 1.4$, so that $\overline{\gamma_i^2} = 2.35$ (Mitrofanov, Ovodova, and Shiyan 1980). For deformations along other axes, Grünisen's coefficients are almost the same. The dashed curve labeled 2 in figure 9 represents the resulting calculated quality factor for longitudinal vibrations in sapphire with a frequency of 38 kHz. This calculated curve agrees much better with the experimental data for temperatures from 20 to 80 K than does the curve (labeled 1) given by equation (3.2).

Using equation (5.5), we can calculate the phonon-phonon induced limit on the value of $Q\omega$ for longitudinal vibrations in sapphire resonators. We find $(Q\omega)_{lim} = 2.4 \times 10^{15}\,sec^{-1}$ at room

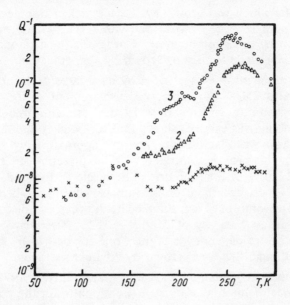

Fig. 10 Comparison of attenuations of longitudinal vibrations in resonators made from sapphire monocrystal (curve 1) and ruby monocrystals (curves 2 and 3) (Bagdasarov, Braginsky, and Mitrofanov 1974).

temperature, and $(Q\omega)_{lim} = 4 \times 10^{15} \text{sec}^{-1}$ at 4.2 K.

Thus there is satisfactory agreement between the calculated and measured values of the sapphire quality factors in the temperature region 20 to 80 K. But at room temperature and liquid-helium temperature, the measured Q-values are lower than the calculated values. It may be that at liquid-helium temperature the Q is actually determined by losses in the suspension system, and at room temperature by defects in the crystal.

Admixtures of different substances are often present in sapphire monocrystals. These admixtures can give rise to relaxation processes that cause additional losses of mechanical energy in the resonator. Bagdasarov, Braginsy, and Mitrofanov (1974) have compared the attenuation of longitudinal vibrations in monocrystals of two types: leucosapphire (with an admixture content of less than $10^{-3}\%$) and ruby (containing about 0.1% Cr_2O_3). Figure 10 shows Q^{-1} as a function of temperature for these experiments. Curve 1 is for a resonator (with $f = 34 \text{kHz}$) made from a leucosapphire

monocrystal which was grown by the method of oriented crystallization. Curve 2 is for a resonator (with $f = 17.3$ kHz) made from
ruby grown by the same method. Curve 3 is for a resonator (with
$f = 13.9$ kHz) made from ruby grown by Verneuil's method.

The attenuation curves for ruby have large peaks at temperatures close to 250 K. An observed displacement of the Q^{-1} temperature maximum with changes in the resonator frequency (the
third harmonic was excited) shows that the peak is due to a relaxation process with activation energy approximately 0.25 eV.
Presumably this peak is associated with the presence of Cr^{3+} ions in
the crystal lattice of the sapphire. The height of the peak decreases
when the concentration of chromium is reduced. When the percentage of chromium is less than 10^{-3}, the peak is not observed at
all to within an accuracy of $\Delta(Q^{-1}) = 10^{-8}$. The character of the
Q^{-1} temperature dependence is the same for both ruby monocrystals, even though they were grown by different methods.

Another experiment (Mitrofanov and Shiyan 1979) reveals
the influence of internal strains in the sapphire on the damping of
vibration. A sapphire resonator was made by the standard method:
a crystal boule was cut and then machined into a cylinder with diamond tools. To avoid cracks and large spalling, all procedures
were carried out as carefully as possible. The crystal surface was
polished and the crystal was annealed at a temperature of 1100°C.
Curve 1 in figure 11 shows the dependence of Q^{-1} on temperature
for such a resonator. The face of the cylinder was then struck sharply, spalling off a piece with an area of 7 mm by 8 mm and a thickness of 1 mm. The damaged crystal was then cooled and the Q of
its vibrations was measured. The measurements revealed a large
peak of losses near 75 K (curve 2 of figure 11). The value of Q^{-1}
was independent of the vibration amplitude for amplitudes in the
range 5×10^{-8} to 10^{-6} cm.

The additional losses could not be associated with the additional dissipation caused by the nick left by the spallation, because
the Q did not change after removal of the layer of crystal that con-
tained the nick. The crystal was then annealed in vacuum at a
temperature of 1100°C. This low-temperature annealing did not
change the temperature dependence of the crystal's Q. After that,
the crystal was annealed at a temperature of 1950°C. Following
this high-temperature annealing, the crystal was ground and polished again, and its Q was measured. The attenuation peak at 75 K
was now completely gone (curve 3 of figure 11). As before, no

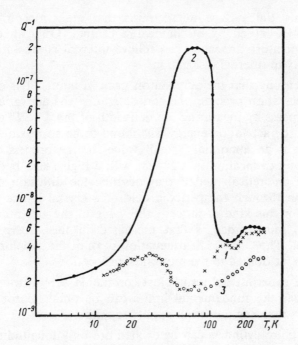

Fig. 11 Influence of spalling in sapphire resonator on Q^{-1}. Curve 1, before spalling; Curve 2, after spalling; Curve 3, after high-temperature firing (Mitrofanov and Shiyan 1979).

dependence of the Q on vibration amplitude was observed.

Presumably the observed peak was associated with a relaxation process. Both the shape of the $Q^{-1}(T)$ curve and the lack of dependence of Q^{-1} on vibration amplitude support this conclusion. This peak is similar to the Bordoni peak (Nowick and Berry 1972) found in metals: (1) after annealing, the peak disappears; and (2) both the height and the location of the peak are independent of the vibration amplitude. It is very probable that the dissipation peak arose as a result of internal strain and dislocations in the crystal caused by the impact. An electron microscope examination has shown that near a point of concentrated load on a sapphire crystal, strain patterns and dislocation lines appear (Belyaev 1974). A low-temperature annealing at 1100°C cannot change the character of the $Q^{-1}(T)$ curve for sapphire because such annealing affects only the surface layers of the crystal, changing their stoichiometry and eliminating the surface microstrains that are concentrated near microcracks caused by mechanical polishing. Only an extended, high-temperature annealing can relieve internal strains and reduce the density of internal dislocations.

Assuming that the attenuation peak in figure 11 is associated with a relaxation process, one can determine the activation energy of this process by measuring the halfwidth of the $Q^{-1}(T)$ curve. In this way, the activation energy was found to be approximately 0.04 eV. This is an abnormally small value for the process of simple motion of a dislocation in a crystal with a high Peirls barrier. But there are theoretical models that describe the diffusion of a kink dislocation through the periodic field of a crystal lattice (Labusch 1965). For this kind of process, the value of the activation energy is of the same order as that measured in the experiment just described. Thus, the 75 K attenuation peak in the sapphire resonator can be attributed to the motion of dislocations.

The experimental results just considered demonstrate that Q-values near the fundamental limits can be obtained for sapphire resonators by using industrial quality sapphire.

Sapphire cylinders can be excited not only longitudinally, but also in other types of modes; for instance, torsional modes. Torsional sapphire resonators have quality factors of $Q \approx 10^8$ at room temperature. Sapphire resonators of noncylindrical shape are used as well. For instance, a tuning fork cut from a sapphire plate has been used to stabilize the frequency of an oscillator (Kochubei and Mitrofanov 1978). This tuning fork was 65 mm long, 14 mm wide, and 7 mm thick. The resonant frequency of the fork was 6.6 kHz. This tuning fork was suspended by tungsten wire 100 μm in diameter in a special chamber. The chamber was then evacuated and placed in a helium dewar. The Q of the tuning fork was equal to 3×10^6 at room temperature and 3×10^7 at 4.2 K. Though these values are high, they are much lower than the fundamental limits for internal friction in sapphire. Losses in the support are crucial for tuning forks (Tomikawa 1979); these losses determined the observed quality factor in this experiment.

Among the materials shown in figure 6, the aluminum alloy Al-5056 and niobium are of special interest. Only recently have they come to be used for fabricating high-Q mechanical resonators. Hirakawa was the first to propose the use of the alloy Al-5056 as a material with a low level of internal losses (Suzuki, Tsubono, and Hirakawa 1978). This alloy is attractive for manufacturing massive resonators; for instance, gravitational-wave antennae weighing several tons. This alloy contains an abnormally large amount of magnesium (about 5%). The magnesium gives the alloy two useful properties, a decrease in dislocation mobility and a decrease in the phonon-electron attenuation (the last partly owing to an increase of electrical resistance). For obtaining the maximum quality factor, the optimum magnesium content is in the 5 to 10% range. But the dependence of the quality factor on the magnesium content is weak in this range, and thus the composition of the alloy Al-5056 is close to optimal (Oide, Tsubono, and Hirakawa 1980).

Niobium has several valuable properties. Because niobium is a superconductor with a comparatively high temperature for transition to the superconducting state and with a high critical value of the magnetic field, niobium can be levitated magnetically. For such levitation the suspension losses can be very small. In addition, acoustic attenuation due to electron-phonon interactions drops sharply at a temperature below the temperature of the superconducting transition. Unfortunately, dislocation damping in niobium is large even at low temperatures (down to 2K) (Blair *et al*. 1980b). Therefore, very strict requirements are imposed on the purity and the treatment of niobium. The quality factors of niobium resonators are very sensitive to the temperature and other conditions under which it is annealed (Blair *et al*. 1980b).

III Electromagnetic Resonators with Small Dissipation

6. Superconducting Cavity Electromagnetic Resonators

Superconducting resonators (SCRs) possess unusually high quality factors and as a result are widely used in modern physics experiments and in engineering. The possibility of making SCRs is due to the phenomenon of superconductivity: below a "critical temperature" T_c (the temperature of the superconducting transition), superconducting materials completely lose their resistance to direct current.

There is a certain amount of resistance to alternating current in superconducting materials, but it is far lower than the resistance of the best conductors in their normal states. As is well known, a static magnetic field penetrates into a superconductor to a depth $\delta \sim 10^{-8}$ to 10^{-7}m, known as the "skin depth." Although the skin depth for a superconductor increases with an increase in the alternating current, it is always less than the skin depth of the normal metal.

Due to this shallow skin depth, the electric field beneath the superconductor's surface has much larger derivatives in the direction normal to the surface than in tangential directions. From this it follows that the electromagnetic field close to the surface of a superconductor can be regarded as a normally propagating plain wave with tangential electric field $\vec{E}_t$ and tangential magnetic field $\vec{H}_t$, which are related to each other by

$$\vec{E}_t = Z_s \vec{H}_t \times \vec{n} , \tag{6.1}$$

(Landau and Lifshitz 1959). Here $\vec{n}$ is the normal vector to the metal's surface. This boundary condition, known as the Shchukin-Leontovich condition, defines the surface impedance Z_s of the superconductor.

For general materials (not necessarily superconductors), the surface impedance is complex, frequency dependent, and related to the complex effective skin depth δ_{sk} by

$$\delta_{sk} = j \, \frac{Z_s(\omega)}{\omega\mu_0} \, . \tag{6.2}$$

Here ω is the angular frequency and μ_0 is the magnetic permeability. The surface impedance $Z_s(\omega)$ can be represented as $Z_s(\omega) = jX_s + R_s$, where R is the active part of the impedance (the resistivity) and X_s is the reactance. At room temperature, whenever Ohm's law $\vec{J} = \sigma\vec{E}$ holds, we have $R_s = X_s = \mu_0\omega\delta_{sk}$. With decreasing temperature, the skin depth drops while the electron mean free path l_e grows. When $\delta_{sk} < l_e$, this leads to an anomalous skin effect.

The theory of the anomalous skin effect (Reoter and Sondheimer 1948; Pippard 1954) shows that in the region $\delta_{sk} < l_e$ the conductivity of a pure metal is finite. This phenomenon is well illustrated by experimental data on the resistance of copper at microwave frequencies: as the temperature decreases from room temperature to liquid-helium temperatures, the surface resistance decreases by a factor of six (Biquard and Septier 1966). By contrast to copper and other normal metals, for superconductors when the temperature drops below T_c the surface resistance drops by 6 to 8 orders of magnitude. Thus one can construct superconducting microwave devices with very small electromagnetic losses.

Surface resistance and residual resistance of superconductors. It is a characteristic feature of superconductivity that the transition to the superconducting state does not occur suddenly at microwave frequencies (by contrast with its sudden onset at zero frequency [DC]). The typical dependence of the surface resistance R_s on temperature below the superconducting transition temperature is shown in figure 12. For $\omega = 0$ and $T < T_c$, the resistance vanishes completely. For nonzero frequencies, the transition to the superconducting state is not as sudden. If the frequency is high enough, there is surface resistance even at $T = 0$. The surface resistance at $T < T_c$ approaches the resistance of the metal in its normal state

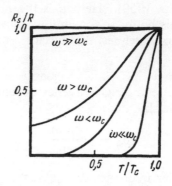

Fig. 12 Temperature dependence of surface resistance R_s of superconductor for various frequencies, compared with surface resistance R of same metal in normal state near T_c (Abrikosov, Gorkov, and Khalatnikov 1958; Abrikosov and Khalatnikov 1959; Mattis and Bardeen 1958).

for frequencies $\omega \gg \omega_c = 2\Delta(0)/\hbar$, where $\Delta(0)$ is the halfwidth of the energy gap of the microscopic theory ($2\Delta(0) = 3.52kT_c$ at $T = 0$).

The large increase in the $T < T_c$ surface resistance R_s as ω rises through ω_c into the range $\omega \gg \omega_c$ is caused by the direct excitation of electrons through the energy gap, which in turn causes the number of normal electrons to be finite even at $T = 0$. As a rule, microwave frequencies are $\ll \omega_c$ for superconducting materials (Pb, Nb, Nb$_3$Sn). For this frequency range and for $T \lesssim 0.5T_c$, the approximate dependence of surface resistance on temperature and frequency (for $\omega \neq 0$) is given by the formula

$$R_s \approx A\,\frac{\omega^{1.7-2}}{T}\exp\left[-\frac{\Delta(0)}{kT}\right] + R_0 , \qquad (6.3)$$

where A is a constant determined by the properties of the given material and R_0 is that part of the resistance which is independent of temperature. Equation (6.3) describes qualitatively how the surface resistance of the superconductor depends on temperature and frequency. To illustrate this, figure 13 shows the dependence of surface resistance on frequency for lead and niobium at $T = 1.8$ K and $T = 4.2$ K (Turneaure 1972).

A quantitative expression for the temperature-dependent part of a superconductor's surface resistance can be obtained from the microscopic theory's formulae for the surface impedance Z_s and

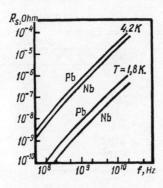

Fig. 13 Frequency dependence of surface resistance for lead and niobium at $T = 4.2\,\mathrm{K}$ and $T = 1.8\,\mathrm{K}$ (Turneaure 1972).

the skin depth δ of the superconductor (Abrikosov, Gorkov, and Khalatnikov 1958; Abrikosov and Khalatnikov 1959). With the help of a function $F(\omega)$, which actually depends on ω, Δ, and T, the quantities δ and Z_s can be put in their simplest form:

$$\delta = \frac{\sqrt{3}}{2}\left[\frac{4m_e \hbar v_F}{3\mu_0 e^2 n_e \Delta}\right]^{1/3} \mathrm{Re}\{F(\omega)\}^{-1/3}, \tag{6.4}$$

$$Z_s = -j\frac{\sqrt{3}}{2}\omega\mu_0\left[\frac{4m_e \hbar v_F}{3\mu_0 e^2 n_e \Delta F(\omega)}\right]^{1/3} = -j2R\left\{\frac{\pi \hbar \omega}{\Delta F(\omega)}\right\}^{1/3}. \tag{6.5}$$

Here R is the surface resistance in the regime of the anomalous skin effect, $\hbar$ is Planck's constant, e is the charge of an electron, n_e is the number density of electrons in the metal, m_e is the electron mass, and v_F is the speed of electrons near the Fermi surface. Equations (6.4) and (6.5), together with the known analytical expression for the function $F(\omega)$, make possible the calculation of the surface resistance at radio and microwave frequencies ($\hbar\omega \ll kT \ll \Delta$):

$$R_s = R\frac{2}{3}\left[\frac{2}{\pi}\right]^{4/3}\left[\frac{\hbar\omega}{2\Delta}\right]^{4/3}\frac{2\Delta}{kT}\ln\frac{4kT}{1.78\hbar\omega}\exp\left\{-\frac{\Delta}{kT}\right\}. \tag{6.6}$$

Unfortunately, these expressions are best suited to explaining the behavior of the surface resistance for type I superconductors, that is, for superconductors in which the skin depth δ is less than the coherence length ξ. Thus, for lead and niobium, which have a

Table 4. Superconductivity Parameters for Lead and Niobium.

Parameter	Material	
	Lead	Niobium
A^*	3	0.97
$\Delta(0)/kT_c$	2.05	1.88
T_c(K)	7.2	9.25
v_F(m/sec)	6×10^5	2.9×10^5
l_e(m)	0.71×10^{-6}	1×10^{-6}
δ(m)	3.08×10^{-8}	3.5×10^{-8}
n_e(m^{-3})	3.7×10^{28}	2.4×10^{28}

skin depth comparable to coherence length, the surface resistance differs from the above formula.

The Mattis-Bardeen theory (Mattis and Bardeen 1958) is in better agreement with experimental data for R_s (Miller 1960; Turneaure 1967; Turneaure and Weissman 1968). However, the good agreement between theory and experiment is obtained only at the price of some modification of the parameters in the analytic expressions that describe the behavior of superconductors. This is why comparisons with experimental values of the surface resistance R_s^{exp} are made using the modified relation (Turneaure and Weissman 1968)

$$R_s^{exp} = A^*R_s + R_0 , \qquad (6.7)$$

where R_s is the theoretically based expression (6.6), A^* is a correction factor, and R_0 is the residual resistance. Moreover, in comparing with experiment, the quantity $\Delta(0)$ in equation (6.6) is adopted as a free parameter. By suitable choices of A^*, $\Delta(0)$, and R_0, a satisfactory agreement between theory and experiment can be achieved.

Comparisons between experiment and theory based on equations (6.6) and (6.7) have been reported for lead and niobium in several different works (Miller 1960; Turneaure and Weissman 1968; Hahn, Halama, and Foster 1968). Table 4 gives the correction parameters and other data which are necessary for the calculation of R_s for lead and niobium. Figure 14 shows the surface resistance as a function of temperature for niobium at a frequency of 3.6 GHz. The measurements were made with superconducting resonators. At low temperatures ($T/T_c \ll 1$), the measured surface resistance differs greatly from its theoretical value [equation (6.7)

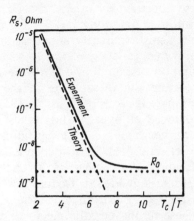

Fig. 14 Temperature dependence of surface resistance for niobium at frequency of 3.6 GHz (Turneaure and Weissman 1968).

with $R_0 = 0$].

In the next few paragraphs we shall consider the main reasons for the appearance of the residual resistance R_0 in superconductors. A residual resistance is the main factor that limits reductions of the surface resistance as $T \to 0$ and makes it impossible to obtain an arbitrarily high quality factor for superconducting resonators.

Various electromagnetic loss mechanisms contribute to the residual resistance R_0, but a full explanation of it has not yet been achieved. Typically, the dependence of R_0 on ω at microwave frequencies is proportional to ω^γ, with $0 \leqslant \gamma \leqslant 2$.

The residual resistance has been studied in detail in several works (Halbritter 1975; Pierce 1974a; Hartwig 1973). The combined influence of different loss mechanisms complicates the investigations and prevents the association of R_0 with any single mechanism. However, it has been established that the residual resistance is reduced when there is little surface contamination and the surface is smooth. Magnetic flux trapped during the transition into the superconducting state can be one of the causes of residual resistance (Pierce 1973). Such flux creates small areas of normal conductivity that produce an anomalous skin-effect and cause a finite resistance at $T = 0$; therefore, $R_0 \neq 0$.

In addition to these effects, mechanisms associated with the electromagnetic generation of phonons have been proposed (Passow 1972). These mechanisms apparently entail residual losses that are essentially nonremovable in superconductors. These losses are caused by the excitation of phonons due to the coupling between the microwave field and the metal ions in the skin layer. The specific focus of the investigations of Passow (1972) and Halbritter (1971) was the influence of hypersound waves generated by the interaction between metal ions and longitudinal and transverse components of the electric field. The dependence of the residual resistance on the frequency in these cases is of the form $R_{0\parallel} \propto \omega^2$ and $R_{0\perp} \propto \omega^0$. Estimates made at frequencies on the order of 10 GHz give residual resistances $R_{0\parallel} \sim 10^{-10}$ ohm and $R_{0\perp} \sim 10^{-8}$ ohm. These results permit one to understand observed differences between the values of R_0 for different oscillation modes.

The lowest experimental values obtained for residual resistance are $R_0 \approx 3 \times 10^{-8}$ ohm for lead (Pierce 1973) and $R_0 \approx (0.5 \text{ to } 2) \times 10^{-9}$ ohm for niobium (Turneaure and Viet 1970; Allen et al. 1971; Kneisel, Stoltz, and Halbritter 1972, Benaroya et al. 1975). These values are unique and barely reproducible. More typically, the values obtained for the residual resistance of lead and niobium are an order of magnitude worse and depend on the purity of the material and the type of surface treatment.

Other kinds of superconducting materials (intermetallic compounds and alloys) were recently found to be helpful in fabricating superconductors with a high value of T_c (Isagawa et al. 1974; Hillenbrand and Martens 1976; Kneisel et al. 1977; Arnolds and Proch 1977; Kneisel, Stoltz, and Halbritter 1979). Their higher values of T_c make it possible to get a lower value for R_s than that obtained with lead and niobium at the same working temperature. For instance, for Nb_3Sn ($T_c \approx 18.2$ K) the surface resistance is $\approx 1.5 \times 10^{-8}$ ohm at a frequency of 3 GHz and a temperature of 4.2 K, whereas the surface resistance for niobium is $\approx 3 \times 10^{-6}$ ohm under the same conditions. But thus far, the experimental value of R_0 for Nb_3Sn at 3 GHz is $\approx (1 \text{ to } 2) \times 10^{-7}$ ohm, a comparatively large value and one greater than the residual resistance for lead and niobium at the same frequency. However, at 9.6 GHz and 4.2 K, $R_s^{exp} \approx 2.9 \times 10^{-7}$ ohm for Nb_3Sn. This value is 70 times lower than the theoretical value of $R_s \approx 2 \times 10^{-5}$ ohm for pure niobium at the same temperature (Kneisel, Stoltz, and Halbritter 1979; Hillenbrand et al. 1977). Therefore, Nb_3Sn can be quite useful for the fabrication of high-quality superconducting resonators.

Some nonlinear effects that cause greater superconductor surface resistance can be attributed to the large quality factors of superconducting resonators, which result in very strong electromagnetic fields at microwave frequencies. These effects are partially associated with either a decrease in the energy gap and an increase in the skin depth, or the appearance of emission currents excited by the electric field.

The Ginzburg-Landau theory (Ginzburg and Landau 1950) enables us to take into account the influence of a microwave-frequency magnetic field on the surface resistance. If $H < H_c$, where H_c is the critical magnetic field, one can approximate

$$R_s(H) = R_s(0)\exp\left\{\frac{1}{2}\frac{\Delta(0)}{kT}\frac{H^2}{H_c^2}\right\}. \tag{6.8}$$

For $H \approx H_c$, this formula says that the ratio $R_s(H_c)/R_s(0)$ at temperature $T = 1.85\,\mathrm{K}$ is equal to 1.8 for lead and 3.3 for niobium (Schwettman et al. 1967). In addition to increasing the surface resistance, a strong electromagnetic field can lead to partial or full destruction of superconductivity due to thermomagnetic effects or the emission of electrons from the surface of the superconductor (Dammertz, Hahn, and Halbritter 1971; Halama 1971). The electromagnetic power dissipated per unit area is $\tilde{W} \approx R_s H^2/2$. Since the thermal conductivity κ and the thickness of the superconductor d_0 are finite, the surface temperature increases to $\Delta T \approx \tilde{W}/d_0\kappa$. The superconductivity can be destroyed if the local surface heats up to temperatures $T \approx T_c$.

A steep rise in surface resistance can also be caused by field emission of electrons induced by the normal component of the electrical field E_n. As a rule, when a superconducting resonator is loaded by an emission current, the critical breakdown value E_{br} for the normal electric field is much less than that in vacuum. Emission currents can exist and produce an increase in surface resistance even when the microwave-frequency normal electric field is rather weak ($E_{br} \approx 5 \times 10^6\,\mathrm{V/m}$). This effect is sensitive to surface conditions and geometry (Dammertz, Hahn, and Halbritter 1971). Perhaps the value of E_{br} can always be increased sufficiently to make it irrelevant to the resonator's Q, either by choosing an appropriate surface geometry or by covering the surface with a dielectric layer. If so, then the limiting quality factor of the resonator will be determined by the magnetic component of the tangential microwave field in the resonator.

The critical value of the oscillating magnetic field H_c^ω at which R_s increases sharply does not have to be equal to the thermodynamic value H_c of the critical magnetic field for a given superconductor. For type I superconductors the value of H_c^ω can sometimes reach the value H_c. For type II superconductors, which in addition to H_c also have a "lower critical field" $H_{c1} < H_c$, it is often H_{c1} which plays the role of the limiting strength of the microwave field in a given resonator. If extremely high quality factors are not desired, the quantity H_c can sometimes be considered the limiting value of the intensity of the microwave frequency field (Halbritter 1972).

The lower critical value of the magnetic field H_{c1} can be expressed in terms of the Ginzburg-Landau parameter κ of the superconductor and the critical thermodynamic field H_c by

$$H_{c1} = H_c \frac{\ln \kappa + 0.08}{\sqrt{2}\kappa} \tag{6.9}$$

(Abrikosov 1965). To make this formula applicable over the entire temperature range $T < T_c$, the parameter κ must be defined by

$$\kappa(T) = \frac{2\sqrt{2}\mu_0 e}{\hbar} H_c(T)\delta^2(T) \tag{6.10}$$

(Werthamer 1969). From these formulae, one can compute the maximum expected value H_{c1} of the critical microwave-frequency magnetic field H_c^ω at any given temperature for superconductors of the second kind. The maximum magnetic field in the 10 GHz frequency band has been achieved with niobium resonators in the oscillation mode H_{011}. The magnetic flux density has been as great as $B_c^\omega \approx 0.16\,\text{T}$ (Schnitzke et al. 1973). For Nb_3Sn, the value $B_c^\omega \approx 0.106\,\text{T}$ has been observed in the H_{011} mode, while the value $B_c^\omega \approx 0.15\,\text{T}$ has been observed in the E_{010} mode (Hillenbrand et al. 1975).

These experimental results demonstrate that it is possible to construct high-power superconducting resonators, which are vital to the development of particle accelerators that use superconducting microwave-frequency resonators as accelerators (Fairbank and Schwettman 1967).

Quality factors of superconducting resonators. Because of their small surface resistance in the microwave frequency band, superconductors can be used for manufacturing resonators with extremely high quality factors, as great as (4 or 5) × 10^{11}. These values are many orders of magnitude greater than the quality factors of resonators made from normal conductors. The quality factor of a superconducting resonator depends on the electromagnetic oscillation mode, and therefore on the electromagnetic field distribution in the resonator cavity.

Using the expressions for the surface resistance and its relationship to skin depth, one can easily construct an analytic expression for the quality factor of an electromagnetic resonator

$$Q_e = \frac{2}{\delta} \int_V H^2 dV \left[\int_S H_t^2 dS \right]^{-1} . \qquad (6.11)$$

Taking into account the connection between δ and R_s, and disregarding the residual resistance, the expression for the quality factor of a superconducting resonator can be written as

$$Q_e = \frac{\mu_0 \omega}{R_s} \int_V H^2 dV \left[\int_S H_t^2 dS \right]^{-1} = \frac{\Gamma}{R_s} , \qquad (6.12)$$

where Γ is a geometric factor determined by the resonating cavity's geometry and the excited oscillation mode. If the field distribution in the resonator is known, both the geometric factor and the quality factor can be calculated easily for each oscillation mode (see, for example, Remis 1949).

For commonly used cylindrical resonators, the quality factors for H and E types of oscillation are given by

$$Q_{mnl}^H = \frac{\omega \mu_0 r_0}{2R_s} \left[\left(\frac{v_{mn}}{r_0} \right)^2 + \left(\frac{\pi l}{L} \right)^2 \right] \times \left[1 - \left(\frac{m}{v_{mn}} \right)^2 \right]$$

$$\times \left[\left(\frac{v_{mn}}{r_0} \right)^2 + \frac{2r_0}{L} \left(\frac{\pi l}{L} \right)^2 + \frac{m^2 \pi^2 l^2}{v_{mn} L^2} \left(1 - \frac{2r_0}{L} \right) \right]^{-1} , \qquad (6.13)$$

$$Q_{mnl}^E = \frac{\omega \mu_0 r_0}{2R_s} \left[1 + \frac{2r_0}{L} \right]^{-1} \quad \text{for } l \neq 0 ,$$

$$Q_{mnl}^E = \frac{\omega \mu_0 r_0}{2R_s} \left[1 + \frac{r_0}{L} \right]^{-1} \quad \text{for } l = 0 . \qquad (6.14)$$

Table 5. Geometric Factor Γ Governing the Q of an Electromagnetic Cavity Resonator.

Oscillation Mode	H_{011}	H_{111}	E_{010}	E_{011}	E_{111}	E_{110}	E_{012}
Γ (ohms)	780	320	302	270	392	480	373

Here v_{mn} is the nth root of the first derivative of the mth order Bessel function, m is the number of halfwaves along the resonator's circumference, n and l are the numbers of halfwaves along the resonator's radius and axis respectively, and r_0 and L are the radius and the length of the resonator. According to equation (6.12), the maximum value of the quality factor is determined by the maximum value of the ratio $\Gamma(\omega)/R_s(\omega)$.

The values of Γ for several low-frequency oscillation modes of cylindrical resonators with equal diameter and length are given in table 5. Substituting in equation (6.12) the appropriate value of Γ from this table, and calculating R_s with the help of equations (6.6) and (6.7), we can find the theoretical limiting value of the quality factor of a superconducting resonator with a given electromagnetic oscillation mode, frequency f, and temperature T. For instance, if the residual resistance did not affect the value of R_s, we could get a quality factor of $Q_e \sim 10^{15}$ for a Nb$_3$Sn resonator at a temperature of 2 K in the 10 GHz band. Actually, the residual resistance restricts the quality factor to the value $Q_c \equiv \Gamma/R_0$, preventing Q_e from being extremely large.

If the resonator cavity is filled with a dielectric with finite loss tangent tanδ and dielectric constant ϵ', there is an additional reduction of the quality factor due to changes in the geometric factor and to induced dielectric losses. In this case, the resonator quality factor is given by

$$Q_e^{-1} = Q_R^{-1} + Q_d^{-1} ,$$

where Q_d is the quality factor governed solely by the dielectric losses. Since the surface resistance of superconductors is extremely small, even a very thin film on the resonator surface, if it has a large tanδ, can greatly decrease the quality factor of the resonator. These effects are strongest in E-mode resonators with a small geometric factor, and in resonators with a small capacitance gap (Kolesov, Panov, and Patnikov 1976).

Frequency stability of superconducting resonators. The eigen-frequency of a cavity resonator is determined in first order by its dimensions. In the case of infinite wall conductivity, the eigenfrequencies of commonly used cylindrical resonators can be expressed in the form

$$\omega_{mnl} = \frac{1}{\sqrt{\epsilon_0 \mu_0}} \sqrt{\gamma_i^2 + (\pi l/L)^2} \,. \tag{6.15}$$

Here $\gamma_i = \gamma_1 = v_{mn}/r_0$ for H-modes and $\gamma_i = \gamma_2 = \mu_{mn}/r_0$ for E-modes (μ_{mn} is the nth root of the Bessel function of the mth order). Using equation (6.15) we can find expressions for the shift of the eigenfrequency of the resonator due to a change of the resonator's dimensions

$$\frac{1}{\omega} \frac{\partial \omega}{\partial L} = \frac{\pi^2 l^2}{\epsilon_0 \mu_o \omega^2 L^3} \,, \quad \frac{1}{\omega} \frac{\partial \omega}{\partial r_0} = -\frac{\gamma_i^2}{\epsilon_0 \mu_0 \omega^2 r_0} \,. \tag{6.16}$$

We can correct for finite wall conductivity by taking into account the imaginary part of the surface impedance, i.e., the surface reactance X_s (Maxwell 1964)

$$\frac{\Delta \omega}{\omega} = -\frac{X_s}{2 Q_e R_s} = -\frac{1}{2} \frac{X_s}{\mu_0 \omega} \int_S H_t^2 dS \left[\int_V H^2 dV \right]^{-1} \,. \tag{6.17}$$

This expression, which for the sake of convenience can be written in the form $\Delta \omega/\omega = -X_s/2\Gamma$, enables us to determine the frequency drift due to changes in the imaginary part of the superconductor's surface impedance (which itself is a function of the frequency, the magnetic field strength, and the temperature).

In many situations, the high quality factors of superconducting resonators impose the difficult requirement that the frequency of the system be highly stable. For instance, the use of a superconducting resonator as a filter implies that the frequency stability of the resonator must be kept within the range $\Delta \omega/\omega \lesssim 0.1 Q_e^{-1}$. This means that for a quality factor $Q_e \approx 10^{10}$, the relative frequency drift must be as small as 10^{-11}. The requirement for stability is even higher if the resonator is used to stabilize a self-excited oscillator. It is difficult to satisfy this requirement even at the extremely low working temperatures generally used for superconducting resonator systems. Small variations in external parameters (temperature, pressure, the power of the microwave-frequency oscillations) can cause a frequency drift that is larger than the bandwidth of the resonator.

We will consider these effects in some detail because they determine the maximum attainable quasistatic stability of a superconducting resonator system.

The frequency changes can be calculated by means of equations (6.16) and (6.17). Frequency drifts due to temperature changes are very important. Such drifts can be characterized by the quantity $\omega^{-1}\,\partial\omega/\partial T$, which is known as the temperature frequency coefficient.

Temperature-induced frequency variations are caused by variations in the resonator dimensions and by changes in the imaginary part of the surface impedance of the superconductor. If the coefficient of linear expansion at a given temperature is known, then the first effect can be estimated from the formula

$$\left[\frac{1}{\omega}\frac{\partial\omega}{\partial T}\right]_\alpha \approx B\alpha(T)\,, \tag{6.18}$$

where B is a factor determined by the type of resonator. For instance, for cylindrical resonators, this expression takes the form

$$\left[\frac{1}{\omega}\frac{\partial\omega}{\partial T}\right]_\alpha \approx \frac{\alpha(T)}{\epsilon_0\mu_0\omega^2}\left[\left(\frac{\pi l}{L}\right)^2 - \gamma_i^2\right]\,. \tag{6.19}$$

The coefficients of linear expansion for most commonly used metals have been measured, at least over the temperature range 10 to 273 K (Novikov 1974). But for the superconducting metals used in the fabrication of resonators, the experimental data on $\alpha(T)$ are discrepant at temperatures $T < T_c$. For such temperatures, the analytic expression for $\alpha(T)$ given by microscopic theory (Kaplun 1974) can be used. The theoretical expressions for niobium and lead are

$$\alpha(T)_{Nb} = 2.95\times10^{-10}\left[0.124T^3 + 665\exp\left(-\frac{1.44T_c}{T}\right)\right]K^{-1}\,,$$

$$\alpha(T)_{Pb} = 11.9\times10^{-10}\left[2.3T^3 + 184\exp\left(-\frac{1.44T_c}{T}\right)\right]K^{-1}\,.$$

$$\tag{6.20}$$

Figure 15 shows $\alpha(T)$, as given by equation (6.20), for niobium and lead. For comparison, the $\alpha(T)$ curve for copper over the range

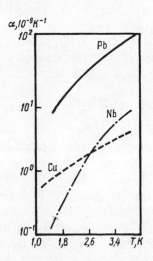

Fig. 15 Temperature dependence of linear expansion coefficients for copper, lead, and niobium (Kaplun 1974).

1.5 to 4.2 K is also given in figure 15. Substituting into equation (6.16) the values of $\alpha(T)$ for niobium and lead at 2 K, we obtain estimates for frequency drifts due to thermally induced changes in the linear dimensions of resonators

$$\left(\frac{1}{\omega}\frac{\partial\omega}{\partial T}\right)_{\alpha\mathrm{Nb}} \approx -5 \times 10^{-10}\ K^{-1}\,,$$

$$\left(\frac{1}{\omega}\frac{\partial\omega}{\partial T}\right)_{\alpha\mathrm{Pb}} \approx -2 \times 10^{-8}\ K^{-1}\,.$$

Thus, to attain a fractional frequency stability of $\Delta\omega/\omega \leq 10^{-15}$ for a niobium resonator, one must hold the temperature constant to within $\sim 2 \times 10^{-6} K$.

The resonant frequency drift due to variations in the imaginary part of the surface impedance of a superconductor X_s can be estimated using formulae that follow from equation (6.5) or using expressions given by Mattis and Bardeen (1958). For approximate calculations it is convenient to use a simple expression for X_s over the temperature range $\hbar\omega < kT < \Delta(0)$

$$X_s = \omega\mu_0\delta(0)\left[1 + C\exp\left(-\frac{\Delta(0)}{kT}\right)\right]. \tag{6.21}$$

Here C is a dimensionless factor determined either experimentally or from the more precise theory. From equation (6.21) and preceding formulae it is not hard to compute the frequency drift of a superconducting resonator due to temperature variations of the imaginary part of its surface impedance:

$$\left[\frac{1}{\omega}\frac{\partial\omega}{\partial T}\right]_{X_s} \approx -C\,\frac{\omega\mu_0\delta(0)}{2\Gamma}\,\frac{\Delta(0)}{kT^2}\,\exp\left[-\frac{\Delta(0)}{kT}\right]. \qquad (6.22)$$

This expression describes the frequency shifts of lead and niobium resonators most accurately if one adopts the following parameters: $C_{Pb}=2.9$, $C_{Nb}=1.8$, $T_{cPb}=7.2\,\mathrm{K}$, $T_{cNb}=9.25\,\mathrm{K}$, $\Delta_{Pb}(0)=2.05\,kT_c$, $\Delta_{Nb}(0)=1.8\,kT_c$, $\delta_{Pb}(0)=4.4\times10^{-8}\mathrm{m}$, and $\delta_{Nb}(0)=4.7\times10^{-8}\mathrm{m}$. For lead and niobium resonators with a geometric factor $\Gamma\approx300\,\mathrm{ohm}$, operating in the E_{010} mode in the 10 m wave band at a temperature of 2 K, equation (6.22) gives the following values for the temperature frequency coefficient

$$\left[\frac{1}{\omega}\frac{\partial\omega}{\partial T}\right]_{X_s\,Pb} \approx -2\times10^{-8}\,K^{-1},$$

$$\left[\frac{1}{\omega}\frac{\partial\omega}{\partial T}\right]_{X_s\,Nb} \approx -3.5\times10^{-9}\,K^{-1}.$$

Thus the frequency drift due to temperature changes of X_s is much greater than the drift due to temperature-induced changes in the cavity dimensions. This is true even for resonators made of niobium, which has a relatively high critical temperature.

However, it should be noted that because of the exponential factor in equation (6.22), the frequency drift caused by changes in the imaginary part of the surface impedance decreases faster with falling temperature than does the frequency drift due to dimension changes. This means that the linear expansion coefficient will be the dominant source of frequency drift below some temperature that is characteristic of the superconducting material used.

As an example, figure 16 shows the dependence of the temperature frequency coefficient on temperature for a cylindrical resonator made from a single piece of niobium and excited at its eigenfrequency of 8.6 GHz in an E_{010} mode (Stein and Turneaure 1978). Above 1.2 K, the value of the coefficient is mainly determined by $X_s(T)$; only below 1.2 K does that part of $\alpha(T)$ produced by the crystal lattice play a decisive role. The electrons'

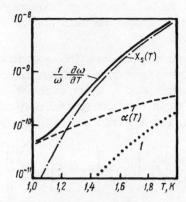

Fig. 16 Temperature dependence of temperature frequency coefficient for niobium resonator with eigenfrequency of 8.6 GHz. Curve labeled 1 is contribution of electron part of niobium's heat capacity (Stein and Turneaure 1978).

contribution to the linear expansion coefficient of niobium (which is related to the electrons' heat capacity) is negligible at $T < 0.5T_c$, and the corresponding influence on $\omega^{-1}\partial\omega/\partial T$ is unimportant throughout the superconducting region (curve 1).

A significant change in the eigenfrequency of a superconducting cavity resonator can be caused by an elastic strain of its walls due to either external isotropic pressure or the stress of the electromagnetic field. Taking cylindrical resonators as an example once again, the resonant frequency drift can be found by applying equation (6.16). In the simplest case of isotropic pressure on the outside faces of a cylindrical resonator excited in the E_{010} mode, we find from equation (6.16)

$$\frac{1}{\omega}\frac{\partial\omega}{\partial P} \approx \frac{S}{Kr_0}.$$

Here K is the mechanical spring constant of the wall, P is the applied pressure, and S is the face area. For resonators made from lead or niobium, even with wall thickness $\sim0.3\lambda$ (where λ is the electromagnetic wavelength), we obtain

$$\frac{1}{\omega}\frac{\partial\omega}{\partial P} \approx 1 \times 10^{-8}\ \text{Torr}^{-1}.$$

This result implies that to achieve high frequency stability one must place the resonator in a high vacuum and must provide thermal contacts for heat removal.

Along with isotropic pressure, the resonator is affected by the pressure of the microwave-frequency electromagnetic field, which is distributed in a complicated manner over the resonator's surface. For instance, the mean electromagnetic pressure in a cylindrical resonator excited in the E_{010} mode is a function of distance from the axis

$$P(r) = \frac{Q_e W}{4\omega V}\left[J_1{}^2\left(\beta_{01}\frac{r}{r_0}\right) - J_0{}^2\left(\beta_{01}\frac{r}{r_0}\right)\right]J_1{}^{-2}(\beta_{01}) , \quad (6.23)$$

where J_0 and J_1 are Bessel functions, W is the total power in the resonator's electromagnetic oscillations, β_{01} is the smallest root of the Bessel function J_0, and V is the resonator volume.

Variations of the pressure [equation (6.23)] due to changes in the electromagnetic power cause frequency drifts. From equation (6.23) one can easily estimate this drift for oscillations in the E_{010} mode

$$\frac{1}{\omega}\frac{\partial\omega}{\partial W} \approx -\frac{Q_e}{2\omega r_0^2 K} . \quad (6.24)$$

For $Q_e \approx 10^9$, $\omega \approx 2 \times 10^{10}\text{sec}^{-1}$, $r_0 \approx 3\,\text{cm}$, and assuming that the mechanical spring constant of the resonator walls is rather large ($K \approx 10^7\text{N/m}$), we find

$$\frac{1}{\omega}\frac{\partial\omega}{\partial W} \approx -3 \times 10^{-6}\,\text{Watt}^{-1} .$$

It is evident that the power pumped into the resonator must be kept very stable.

Frequency drifts due to electromagnetic power changes have been studied by Stein (1975) in high-quality niobium resonators with a geometric factor of $\Gamma = 180\,\text{ohm}$, operating in the E_{010} mode. The fractional frequency drift was proportional to H^2 as expected and had a magnitude $\Delta\omega/\omega \approx 10^{-7}$ at a frequency of 8.6 GHz as H was changed from 1.2×10^4 to 3.6×10^4 At/m. Thus, when a microwave-frequency resonator is used for the measurement of small frequency changes, fluctuations in its electromagnetic power can give rise to large errors.

In addition to its strain-induced influence on the resonant frequency, the electromagnetic pressure can cause an additional frequency shift due to changes in the imaginary part of the surface resistance $X_s(H)$ caused by the microwave-frequency magnetic field. According to the Ginzburg-Landau theory (Ginzburg and Landau 1950), this effect is produced by the dependence of the skin depth δ on the magnetic field strength

$$\delta(H) \approx \delta(0)\left[1 + F(\kappa)\left|\frac{H}{H_c}\right|^2\right].$$ (6.25)

Here $F(\kappa)$ is a function of the Ginzburg-Landau parameter κ, and $\delta(0)$ is the skin depth at $H \rightarrow 0$. Using this expression and taking account of the relationship between the magnetic field strength and resonator power, we can easily obtain an analytic expression for the resonant-frequency drift caused by power-induced changes of the skin depth:

$$\frac{1}{\omega}\frac{\partial\omega}{\partial W} \approx -\frac{2\delta(0)Q_e F(\kappa)}{\Gamma V H_c^2}.$$ (6.26)

This formula describes the frequency drift for resonators excited in the E_{010} mode. For excitations in other oscillation modes, this expression changes by only a factor of ≤ 3, which depends on the distribution of the magnetic field in the resonator.

As an example, for a niobium resonator ($\delta(0) \approx 5 \times 10^{-8}$m, $\kappa \approx 0.8$) with quality factor $Q_e \approx 10^9$, equation (6.26) gives a resonant frequency drift of

$$\frac{1}{\omega}\frac{\partial\omega}{\partial W} \approx -3 \times 10^{-8} \text{ W}^{-1}.$$

For cavity resonators this effect causes a smaller change of frequency than the strain-induced influence of the electromagnetic power. However, this effect becomes significant for resonators made in a more mechanically rigid fashion (see below).

The most important effects influencing the frequency stability of high-quality superconducting resonators are those just considered. However, many other factors can influence the eigenfrequency of a superconducting resonance system. For instance, fluctuations in the absorption of high-frequency power by the resonator walls, which have a finite thermal conductivity, will change the wall temperature and cause a frequency drift (Biquard, Grivet, and Septier 1968). Frequency drifts can also be caused by penetrating radiation, acoustic disturbances, frozen magnetic flux, and other phenomena (Stein and Turneaure 1972).

A very important requirement for superconducting resonators is that the resonant frequency be reproducible. The frequency reproducibility determines the usefulness of the resonator both as a frequency reference and as the stabilizing element of secondary frequency standards. Unfortunately, the resonant frequency of a superconducting cavity resonator changes significantly on long timescales. Frequency changes are especially large after the resonator is warmed up or cooled down. Typically, the cumulative fractional frequency change is $\sim 10^{-5}$ to 10^{-6} (Kaplun 1974) after a number of heating and cooling cycles. The frequency reproducibility is only $\sim 10^{-7}$ even after a large number of such cycles. Therefore, to get a high long-term frequency stability it is essential to keep the resonator continually in a vacuum at an extremely low temperature, and practical problems make it difficult to do this.

Methods of constructing superconducting resonators. Let us now consider the main methods used to construct superconducting resonators:

High-quality superconducting resonators are made from high purity superconducting materials with high critical magnetic fields H_c and temperatures T_c. These characteristics are necessary in order to get small surface resistances R_s, large high-frequency critical fields H_c^ω, and small frequency drifts due to the temperature dependence of the surface reactance.

The most common materials used in superconducting resonance devices are lead and niobium. These materials are characterized by comparatively high critical magnetic fields and temperatures ($T_{cPb} = 7.2\,\text{K}$, $H_c = 6.4 \times 10^4\,\text{At/m}$; $T_{cNb} = 9.25\,\text{K}$, $H_c = 1.6 \times 10^5\,\text{At/m}$), and techniques of preparing these materials are well developed. Other materials (specifically, intermetallic compounds and alloys) are not very acceptable as superconducting materials at microwave frequencies because of their large residual resistances. However, the intermetallic compound Nb_3Sn ($T_c = 18.2$) has achieved a quality factor of $Q_e \approx 6 \times 10^9$ at a frequency of 9.5 GHz (Hillenbrand *et al.* 1977). Thus, this material holds much promise for future use in microwave-frequency devices.

The purity of the material and the absence of surface defects are crucial conditions for obtaining high quality factors in superconducting resonators. Since the skin depth δ and surface resistance R_s are small, even trace contaminants introduced by careless mechanical treatment may cause a sharp increase in the surface resistance and a reduction of the critical magnetic field strength. These conditions have motivated the development of special procedures for the machining and treatment of superconducting surfaces, and these have resulted in the record high values for the quality factors of microwave-frequency resonators.

Lead resonators are commonly constructed by machining, electroplating, vacuum casting, and evaporative coating (for details see Didenko 1973; Mende, Bondarenko, and Trubitsyn 1976). The machining involves turning the bulk material on a lathe, and stamping, forging, or extruding. Though mechanical methods of manufacture are easier than machining, as a rule they do not give large quality factors (Mende, Bondarenko, and Trubitsyn 1976). But even with careful machining, additional surface treatment is necessary, including chemical polishing. Mechanically fabricated resonators have quality factors of $Q_e \approx 10^8$, and some are slightly higher. These low quality factors might be caused by defects in the surface layers produced by the mechanical fabrication processes.

Vacuum casting makes it possible to fabricate resonators without butts and slits that intersect the lines of current flow. Still, this method does not produce resonators with high quality factors. The electroplating of a layer of lead on a metal base gives much better results (Mende, Bondarenko, and Trubitsyn 1976). This method is more complicated because additional treatment of the metal base is necessary, and the perfection of this treatment largely determines the quality factor. The lead coating is made several microns thick in order to get a smoother external surface. Resonators are assembled and dried in an atmosphere of nitrogen or inert gas since the surface of the working layer of lead oxidizes quickly in air. This technique has made it possible to get a record value of $Q_e \approx 4 \times 10^{10}$ for the quality factor of a lead resonator with an eigenfrequency of 12 GHz (Pierce 1973).

The evaporative coating method also provides high quality factors (up to $Q_e \sim 10^9$). But this process must be carried out in a superhigh vacuum ($P \leq 10^{-7}$ to 10^{-9}Torr) using extremely pure lead. It should be noted that the rate of precipitation of the lead onto the base metal can, depending on the pressure, considerably influence the residual resistance and consequently the attainable quality factors. Table 6 gives the basic parameters of

Table 6. Lead Superconducting Cavity Resonators.

Fabrication method	Oscillation mode	f (GHz)	T (K)	Q_e	Reference	Notes
Machining	H_{011}	9.4	2.0	3.0×10^7	Khaikin 1961	
	H_{111}	3.16	4.2	1.6×10^7	Nguyen Tuong Viet 1967	
Vacuum casting	H_{012}	10.0	4.2	7×10^6	Mende et al. 1976	
Electroplating	H_{011}	2.86	4.2	2×10^8	Pierce et al. 1964	
			1.8	2.5×10^9	Pierce et al. 1964	
	H_{012}	10.0	2.0	2×10^8	Pierce et al. 1964	
	H_{013}	12.2	1.3	4×10^{10}	Pierce 1973	Magnetic screening
Evaporative coating	H_{011}	3.16	1.6	6.5×10^8	Nguyen Tuong Viet and Biquard 1966	Magnetic screening
	H_{011}	2.64	2.0	1.5×10^9	Flecher et al. 1969	Magnetic screening
(on sapphire)	E_{010} ($\Gamma \sim 70$ohm)	2.87	2	3.5×10^8	Bagdasarov, Braginsky, and Zubietov 1977; Braginsky and Panov 1979	Magnetic screening

superconducting resonators made from lead by various methods. Unfortunately, even in a good vacuum the surface of a lead resonator is soon oxidized, and therefore, lead resonators are not widely used in microwave technology.

Niobium resonators are widely used at microwave frequencies. Niobium has several advantages over lead: a higher critical temperature T_c, a higher critical magnetic field H_c^ω, and less tendency to oxidize. Additionally, niobium resonators have greater mechanical rigidity than lead resonators, permitting their use as self-excited oscillators with a higher frequency stability. Three types of niobium are used In the fabrication of resonators: totally recrystallized niobium, niobium assembled by an electric arc, and niobium assembled by electron beam welding (Weissman and Turneaure 1968). There are several methods of niobium resonator fabrication: small resonators as a rule are fabricated from a single piece of bulk niobium. Thus, it is not necessary to weld separate parts of the resonator together with an electron beam or join them with lead or indium spacers. Large resonators (used in accelerators) are extruded from sheet niobium (Padamsee *et al.* 1976). Niobium resonators are also fabricated by depositing a thin niobium layer on an internal metal backing, e.g., by evaporative coating in a vacuum. The inner backing is eliminated later, after the construction of a thick shell of some other material on the niobium layer's outer face. This shell forms the stable body of the resonator (Meyerhoff 1969).

To assure high quality factors, niobium resonators undergo a treatment after fabrication that includes annealing in a superhigh vacuum for several hours at a temperature of 1800 to 1900°C, electrochemical polishing, and anodizing oxidation (used to create a protective dielectric film of Nb_2O_5). There are two special treatment procedures that allow niobium resonators to reach quality factors of order $Q_e \approx 10^{10}$ to 10^{11}. In the first procedure, the resonator is annealed after mechanical preparation in a superhigh vacuum at a pressure of $P \approx 10^{-9}$ Torr for 2 to 4 hours; then it is chemically polished with a mixture of nitric and hydrofluoric acid at a temperature of 0°C; and then it is carefully washed and annealed a second time in a superhigh vacuum (Turneaure and Viet 1970). The second procedure is used for resonators that have been fabricated from niobium welded with an electron beam or an electric arc. After fabrication, the resonator is chemically polished in a mixture of sulfuric and hydrofluoric acid. Deformed layers

100 to 200 μm thick are eliminated in this process. Then the resonator undergoes anodizing oxidation in a solution of ammonia. This process leaves a dielectric film of Nb_2O_5 that is approximately 0.3 to 0.5 μm thick (Diepers et al. 1971; Diepers and Martens 1972). Variations of these two procedures are also used. For instance, if necessary chemical polishing can be replaced by electrochemical polishing, and oxypolishing can replace a single anodising oxidation. In the process of oxypolishing, a film of Nb_2O_5 is formed and cleaned many times.

These procedures have made it possible to obtain quality factors for niobium superconducting resonators as high as $Q_e \approx (1 \text{ to } 5) \times 10^{11}$ (Turneaure and Viet 1970; Allen et al. 1971; Kneisel, Stoltz, and Halbritter 1972). Also, anodizing oxidation enables the value of the critical magnetic field in the resonator (H_c^ω) to approach the limiting value (H_{c1}) (Schnitzke et al. 1973). As a rule, the record values for the quality factor listed above were obtained with resonators fabricated from a single piece of extremely pure niobium. Other methods of manufacturing niobium resonators, though more economical and convenient, do not provide such excellent results. For example, Padamsee et al. (1976) constructed sectional resonators in the 10 cm microwave band from extruded sheet niobium. After annealing these resonators at a temperature of 1900°C for 12 hours and subsequently polishing them electrochemically, the maximum quality factor obtained was $Q_e \approx 1.4 \times 10^{10}$.

Another method of fabrication has yielded a quality factor of $Q_e \approx 2 \times 10^{10}$ at a frequency of 11.2 GHz in the H_{011} mode. This method involved the galvanic deposition of a niobium layer onto a copper backing, subsequent dissolution of the backing, and annealing of the resulting niobium mold (Meyerhoff 1969). For niobium resonators fabricated by evaporative coating in a vacuum, the quality factors were still lower. Although this method is quite simple, admixtures of other gases interacting with the vaporized niobium prevent the formation of a niobium film with a low residual resistance. The quality factors of the best resonators made by this method were $Q_e \approx 6 \times 10^8$ in the 3 GHz frequency range with a film thickness of 1 to 2 μm. The quality factors increased with the thickness of the film to a maximum value of 3×10^9 for a film that was 125 μm thick (Padamsee et al. 1976). It should be noted that when coating the inner surface of a metal backing, the upper layers of the coating material are the most important; and for practical reasons these layers are the most pure. That is why an increase in film thickness causes a reduction of the residual resistance to its minimum value R_{0min} at some thickness of the coating layer, and then further thickness increases produce no further reduction of resistance.

Table 7 gives data on experimental results for the quality factors of niobium resonators with weak electromagnetic fields ($H < H_c^\omega$). The highest quality factor that has been achieved is 5×10^{11} at a frequency of 10.5 GHz (Allen *et al.* 1971). It is worth noting that for this quality factor and frequency the relaxation time for the resonator's electromagnetic oscillations is ≈ 14 sec.

The maximum strength of the microwave-frequency electromagnetic field is an important characteristic of resonators. Table 8 gives the parameters of niobium resonators with fields close to H_c^ω. From these data, one can find the maximum power

$$W = \frac{\omega}{Q_e} \frac{\mu_0}{2} \int_V (H_c^\omega)^2 dV \qquad (6.27)$$

deliverable to a resonator without debilitating its performance.

Intermetallic compounds and alloys are not typically used for the fabrication of superconducting resonators. The results obtained from their occasional use are worse than results obtained from niobium. However, even these results show that intermetallic compounds and alloys are somewhat promising for applications in cryogenic microwave equipment. For example, table 9 gives the characteristics of resonators made from superconducting alloys and compounds.

In conclusion, we note that superconducting cavity resonators with high quality factors and high critical magnetic fields have made it possible to improve considerably the performance of many AC cryogenic devices and to improve significantly the precision of several methods of physical measurement.

7. Properties of Superconducting Resonators with Dielectric Interiors

In section 6 we discussed the high-frequency properties of superconducting materials and the features of superconducting cavity resonators. The quality factors of these resonators are

Table 7. Niobium Superconducting Cavity Resonators, Weakly Excited.

Fabrication method	Oscillation mode	f (GHz)	T (K)	Q_e	Reference	Notes[*]
From bulk niobium treated by	H_{011}	10.5	1.3	5×10^{11}	Allen et al. 1971	HVA+ChP
	H_{011}	≈ 3	1.2	4×10^{11}	Kneisel et al. 1972	EP+AO
electron-beam fusion	E_{010}	8.6	1.25	$>1 \times 10^{11}$	Turneaure and Viet 1970	HVA+ChP
	H_{011}	9.5	1.45	3×10^{10}	Diepers and Martens 1972	EP+AO
Extrusion	E_{110}	3	1.4	1.4×10^{10}	Padamsee et al. 1976	HVA+ChP
	E_{010}	3	1.4	1×10^{10}	Klein et al. 1980	HVA
Electroplating	H_{011}	11.2	1.2	2×10^{10}	Meyerhoff 1969	HVA

[*]HVA - High Vacuum Annealing; ChP - Chemical Polishing; EP - Electrolitic Polishing; AO - Anodic Oxidation

Table 8. Niobium Superconducting Cavity Resonators with Fields Close to H_c^ω.

Oscillation mode	f (GHz)	H_c^ω (At/m)	E_c^ω (MV/m)	$Q_e(H_c^\omega)$	Reference
H_{011}	9.5	1.27×10^5	-	1×10^{10}	Schnitzke *et al.* 1973
H_{011}	10.5	1×10^5	-	2×10^{10}	Diepers & Martens 1972
E_{010}	8.6	8.6×10^4	70	1×10^{10}	Turneaure & Viet 1970
Helical resonator	0.092	9.5×10^4	37	2.4×10^8 $Q_e(0) = 9.4 \times 10^9$	Benaroya *et al.* 1975
E_{010}	4.5	4.6×10^4	-	2×10^9	Lagomarsino *et al.* 1979
E_{110}	3	2.4×10^4	-	1×10^{10}	Padamsee *et al.* 1976

Table 9. Superconducting Cavity Resonators Made from Intermetallic Compounds and Alloys.

Material	Oscillation mode	f (GHz)	T (K)	Q_e	T_c (K)	Reference
Nb_3Sn	H_{011}	9.5	1.5	6×10^9		Hillenbrand *et al.* 1977
	H_{011}	9.5	4.2	2.7×10^9	18.2	Hillenbrand *et al.* 1977
$Nb_{0.4}Ti_{0.6}$	H_{011}	3.9	1.5	2.9×10^9	9.8	Giordano *et al.* 1975
$Mo_{0.75}Re_{0.25}$	H_{011}	11.3	1.8	1.4×10^9	10.5	Agyenman *et al.* 1977

comparable to the quality factors of quantum radiative transitions in absorption cells (Bagaev, Baklanov, and Chebotaev 1972). Unfortunately, however, the long-term frequency stability and reproducibility of superconducting cavity resonators are poor, and this makes superconducting resonators impractical for devices that demand long-term frequency stability (e.g., highly stable self-excited oscillators).

These limitations of cavity resonators can be reduced considerably by making the interiors of the resonators from dielectric materials with small dielectric losses and covering them with a superconducting film (figure 17; Bagdasarov, Braginsky, and Zubietov 1977; Braginsky and Panov 1979; Balalykin, Zubietov, and Panov 1978). In such resonators the superconducting surface that interacts with the electromagnetic field is shielded from atmospheric degradation, and the mechanical rigidity of the dielectric can significantly improve the mechanical stability of the resonator. One can hope thereby to get better reproducibility of the eigenfrequency and quality factor as well as greater frequency stability.

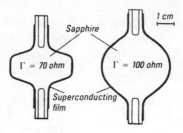

Fig. 17 General forms of superconducting resonators with dielectric interiors (Bagdasarov, Braginsky, and Zubietov 1977; Braginsky and Panov 1979; Braginsky, Panov, and Vasiliev 1981).

For the construction of such a resonator one needs a dielectric with suitably high levels of mechanical rigidity and low levels of dielectric losses. The dielectric must also have properties that lead to a small temperature frequency coefficient; specifically, it must have a small thermal expansion coefficient and a dielectric constant that is a slowly varying function of temperature, and it must be as rigid and firm as possible. Sapphire is the dielectric that best meets all these requirements.

Sapphire is, all in all, one of the best dielectrics. Its loss tangent is small at low temperatures, $\tan\delta \leq 10^{-9}$ (Braginsky, Panov, and Vasiliev 1981), and its linear expansion coefficient is likewise small, $\alpha(T) \approx 6 \times 10^{-13} T^3$ K^{-1} (Vasiliev and Panov 1980). Its dielectric constant ϵ' is a slowly varying function of direction and equals 8.6 and 10.55 for the two orthogonal axes in the crystal. In microwave devices, sapphire is often used as a substrate for microstrip lines and for the fabrication of superconducting microstrip resonators (Di Nardo, Smith, and Abrams 1971). But the quality factors determined by radiative losses and the surface resistance of the superconductor are not more than $Q_e \approx 10^6$ for strip resonators.

In cavity resonators, radiative losses are practically absent and the quality factors are given by

$$\frac{1}{Q_e} = \frac{R_S}{\Gamma} + \tan\delta + \chi'' + \frac{\varepsilon_{ac}}{\varepsilon_0} , \qquad (7.1)$$

where Γ is the geometric factor of the resonator, χ'' is the fractional energy loss induced by paramagnetic impurities in the dielectric,

Table 10. Surface-resistance Parameters for Superconductors.

Material	$\Delta^*(0)$	T_c (K)	B^*
Nb	1.88	9.25	7.1×10^{-22}
Pb	2.05	7.2	5.2×10^{-22}
Nb$_3$Sn	2.1	18.2	5.0×10^{-22}

$\mathcal{E}_{ac}$ is the electromagnetic energy transmitted into acoustic vibrations per cycle of oscillation, and $\mathcal{E}_0$ is the total electromagnetic energy stored in the resonator. To assess the possibility of obtaining high quality factors, we must estimate the contribution of each term in equation (7.1).

The losses due to the surface resistance of a superconductor can be found for dielectric resonators, as for cavity resonators, by using equations (6.3) to (6.6). It is convenient to express the ratio R_s/Γ at $T < T_c$ in the form

$$\frac{R_S}{\Gamma} \approx \frac{B^*}{\Gamma} \, \omega^{1.7} \exp\left(-\frac{\Delta^*(0)T_c}{T}\right) + \frac{R_0}{\Gamma}, \qquad (7.2)$$

where B^* and $\Delta^*(0)$ are the parameters given in table 10 for Pb, Nb, and Nb$_3$Sn superconductors.

The first term in equation (7.2) yields, as above, the fundamental losses in a given superconductor at a given frequency and temperature. The second term gives the residual losses. To illustrate, consider a dielectric resonator whose quality factor is not limited by the residual resistance R_0 or any other losses except fundamental ones. Suppose further that the dielectric resonator is covered by a Nb$_3$Sn film and possesses the geometric factor $\Gamma \approx 10^2$ohm at a frequency of $\omega = 2 \times 10^{10}$sec^{-1}. From equations (7.1) and (7.2) and table 10 we then find that at 2 K the quality factor can be as large as 10^{15}, which is much larger than any quality factors ever attained for either cavity resonators or quantum radiative transitions. This remarkable, idealized performance arises because for $T \to 0$ and $\omega \ll \Delta(0)/\hbar$, the surface resistance due to fundamental processes goes to zero. The appearance of a residual resistance R_0 in the superconductor is not fundamental; rather, it is caused by practical problems.

Dielectric losses are perhaps the main obstacle to achieving high quality factors for dielectric resonators. There are only a few dielectrics with sufficiently small losses in the microwave frequency range. The dielectric losses can be determined by the loss tangent $\tan\delta - \epsilon''/\epsilon'$, where ϵ'' and ϵ' are the imaginary and the real parts of the complex dielectric constant for the given material. These losses are associated either with the process that gives rise to polarization in the dielectric or with the dielectric's conductivity σ. The contribution of the conductivity can be estimated as (Skanavi 1949)

$$\tan\delta_\sigma \approx \frac{4\pi \times 9 \times 10^{13}\sigma}{\omega\epsilon'} .$$ (7.3)

For instance, at $T < T_D$ (T_D is the Debye temperature) we have for leucosapphire $\sigma \approx 10^{-18}$ to 10^{-19}mho/m (Belyaeva 1974; Harrop and Creumer 1963), so that at a frequency of $\omega - 2 \times 10^{10}\text{sec}^{-1}$

$$\tan\delta_\sigma \lesssim 10^{-14} \text{ to } 10^{-15} .$$

During dielectric relaxation, losses are associated with the electromagnetic disturbance of phonons in the dielectric. The relaxation process is accompanied by dissipation of the energy in the electromagnetic field. The losses are dependent on the type of crystal involved, and the temperature dependence of the losses is very sensitive to the symmetry of the crystal. Dielectric losses in crystals with a center of symmetry are much smaller than losses in crystals without such a center.

V. Gurevich (1979, 1980) has developed a theoretical approach that permits one to obtain analytical formulae for the $\tan\delta$ of monocrystals with various symmetries. His approach is based on the assumption that for an ideal monocrystal the anharmonicity of the lattice is the only source of losses. One of the resulting loss mechanisms is the decay of an electromagnetic photon into a pair of phonons.

Gurevich's formulae permit one to obtain approximate estimates for $\tan\delta$. For example, for a perfect monocrystal of Al_2O_3 the Gurevich estimate is

$$\tan\delta \simeq \frac{\eta\omega(kT)^5}{\epsilon\bar{v}^5(kT_D)^2\hbar^2} ,$$ (7.4)

if $kT >> \hbar\omega$. Here ρ is the density of the crystal, $\bar{v}$ is the mean value of the speed of sound, ϵ_0 is the dielectric constant of sapphire, and η is a dimensionless factor of order 10 to 10^2, which depends on unknown details of the nonlinearity of the sapphire lattice. Substituting into (7.4) $T = 300\,\mathrm{K}$ and $\omega = 2 \times 10^{10}\mathrm{rad/sec}$, we obtain $\tan\delta \simeq 10^{-6}$ to 10^{-7}. In the last part of this section we shall describe the lowest observed values for the $\tan\delta$ of Al_2O_3 at low temperatures; and in section 8 we shall present observational data on $\tan\delta$ for a wide range of temperatures. Those data (see figure 24) agree rather well with the Gurevich estimate (7.4) at $T > 60\,\mathrm{K}$.

As a rule, defects produce significant losses (Vinogradov 1969). An estimate of the influence of defects can only be made for a specific crystal because the specific type and distribution of defects must be known. In specific cases the estimate can be based on the results of Vinogradov's (1969) work. For example, for sapphire with a defect percentage of $< 10^{-3}\%$, an upper limit on the loss tangent due to defects is

$$\tan\delta \leq 10^{-9}$$

at a frequency $\omega \approx 2 \times 10^{10}\mathrm{sec}^{-1}$ and temperature $T \approx 4\,\mathrm{K}$.

Paramagnetic losses are produced by paramagnetic atoms in the dielectric. Resonant absorption is the principal mechanism of paramagnetic losses (Abragam and Bleaney 1970; Veilsteke 1963; Altschuler and Kozyrev 1972). The susceptibility χ'' is determined by the line shape $f(\omega)$ of the resonance

$$\chi'' = \pi\omega\chi_0 f(\omega) , \qquad (7.5)$$

where χ_0 is the static susceptibility. At low temperatures and for small paramagnetic ion concentrations, the line has a minimum width and an almost Lorentzian shape

$$f(\omega) = \frac{1}{\pi} \frac{\Delta\omega_L}{(\Delta\omega_L)^2 + (\omega - \omega_L)^2} . \qquad (7.6)$$

Here ω_L is the transition resonant frequency, and $\Delta\omega_L$ is the linewidth defined in terms of the half life τ^* of the excited state by the expression $\Delta\omega_L \approx (\tau^*)^{-1}$.

Equations (7.5) and (7.6) show that the paramagnetic losses are proportional to the static susceptibility χ_0, which can most easily be estimated for a substance in which the paramagnetic atoms can be treated as weakly coupled dipoles (low impurity concentration). In particular, for weak magnetic fields ($\mu_B g J H \ll kT$, where μ_B is the Bohr magneton, g is the Lande factor, and J is the atom's total angular momentum quantum number), and in cases where the contribution of the nuclear magnetic moment to the atom's moment is small, we have for the static susceptibility (Altschuler and Kozyrev 1972)

$$\chi_0 \approx \frac{N}{3kT} \, \mu_B^2 g^2 J(J + 1) . \tag{7.7}$$

Here N is the concentration of paramagnetic ions. Substituting this expression into equation (7.6), we find for paramagnetic losses

$$\chi'' \approx \frac{\omega N}{3kT} \, \mu_B^2 g^2 J(J + 1) \, \frac{\Delta \omega_L}{(\Delta \omega_L)^2 + (\omega - \omega_L)^2} . \tag{7.8}$$

A pure sapphire crystal, composed of Al^{3+} and O^{2-} ions, possesses no paramagnetic properties. Paramagnetic losses in sapphire can only appear due to the presence of paramagnetic ions; the major contributors are Cr^{3+} ions. If the Cr^{3+} ion concentration is high enough, the sapphire becomes ruby, which has well-known properties (Veilsteke 1963). This makes it possible to estimate the dielectric losses due to Cr^{3+} impurities. For example, for an electromagnetic resonator with a weak magnetic field, we find from equation (7.8) that for sapphire with a chromium concentration of $N \approx 5 \times 10^{17} \text{cm}^{-3}$ ($\sim 10^{-3}\%$) at $T = 2K$, with $g = 2$, $J = 5\frac{1}{2}$, $\Delta \omega_L \approx 10^8 \text{sec}^{-1}$, and $\omega_L \approx 7.3 \times 10^{10} \text{sec}^{-1}$ (Veilsteke 1963), the dielectric loss at frequency $\omega = 2 \times 10^{10} \text{sec}^{-1}$ is

$$\chi'' \lesssim 10^{-9} .$$

Recalling [equation (7.1)] that $Q_e^{-1} = \chi'' +$ (other contributions), we see that even very small paramagnetic impurities in a sapphire crystal can lead to energy dissipation that lowers the resonator quality factor. This effect must be taken into account when a superconducting resonator is filled with a dielectric.

Let us turn to the analysis of acoustic losses $\varepsilon_{ac}/\varepsilon_0$. Acoustic losses arise due to the coupling of energy from microwave-frequency electromagnetic oscillations into acoustic vibrations (hypersound). By contrast with the phonon mechanism of residual losses (Passow 1972), hypersound waves arise in the walls of a dielectric resonator independently of the scale of inhomogeneities in the superconducting film. Hypersonic losses are driven by the electromagnetic pressure P on the walls of the resonator. There are static and alternating components of the pressure, and the latter have twice the frequency of the electromagnetic field

$$P = \frac{\mu_0 H_t^2 - \epsilon' \epsilon_0 E_n^2}{2} + \frac{\mu_0 H_t^2 + \epsilon' \epsilon_0 E_n^2}{2} \sin 2\omega t , \tag{7.9}$$

where E_n is the normal component of the microwave-frequency electric field.

The static component does not cause losses, although it deforms the resonator and shifts its eigenfrequency. The losses are associated with the alternating component acting on the superconducting walls, which in turn generate longitudinal hypersound waves that propagate into the volume of the dielectric carrying a stationary flux of elastic strain energy

$$U = \int_S v\rho_S dS , \qquad (7.10)$$

where U is the total power, v is the sound velocity, ρ_S is the density of elastic strain energy, and S is the surface area of the resonator. Assuming that the hypersound-wave propagation in the volume has a diffusive character, and using the expression $\rho_S = P^2/2Y$ for the density of elastic strain energy, we can estimate the hypersonic losses in the dielectric electromagnetic resonator:

$$\frac{\varepsilon_{ac}}{\varepsilon_0} \approx \frac{1}{\omega} \left(\frac{\pi\varepsilon_0 v}{2Y} \right)^{1/2} \left[\int_S (\mu_0 H_t^2 + \epsilon_0\epsilon' E_n^2)^2 dS \right]^{1/2} \qquad (7.11)$$

$$\times \left[\int_V (\mu_0 H^2 + \epsilon_0\epsilon' E^2) dV \right]^{-1} . \qquad (7.11)$$

A numerical estimate can be performed for a specific distribution of the electromagnetic field in the resonator. For example, for the E_{010} oscillation mode one obtains

$$\frac{\varepsilon_{ac}}{\varepsilon_0} \approx \frac{1}{\omega V} \left(\frac{\pi\varepsilon_o v S}{2Y} \right)^{1/2} . \qquad (7.12)$$

For a sapphire resonator with $Y = 4 \times 10^{11}$Pa, $v = 10^4$m/s, $V = 10^{-3}$m^3, $S = 1.5 \times 10^{-3}$m^2, $\omega = 2 \times 10^{10}$sec^{-1}, and $\varepsilon_0 = 10^{-3}$W, equation (7.12) gives

$$\varepsilon_{ac}/\varepsilon_0 \approx 1 \times 10^{-12} .$$

It should be noted that according to equation (7.12), the hypersonic losses depend on the electromagnetic energy ε_0 stored in the resonator and increase as $\varepsilon_0^{1/2}$.

In view of these estimates for a superconducting resonator with a sapphire interior, it should be possible to attain quality factors for dielectric resonators that are comparable to those of superconducting cavity resonators. However, we must point out that the presence of the dielectric in the resonator's interior reduces its geometric factor Γ by approximately $\sqrt{\epsilon'}$, which decreases its quality factor by the same amount in comparison with a cavity resonator that is excited in the same oscillation mode and has the same surface resistance. This implies that the use of a dielectric with a small or moderate dielectric constant is preferable, other things being equal.

Another important feature of superconducting resonators with dielectric interiors is the stability of their eigenfrequency. Frequency drift is caused by many factors such as temperature, pressure, strengths of the E and H fields in the resonator's interior, and so on. Nevertheless, all frequency changes are due most directly to changes in the imaginary part of the surface resistance, to changes in the dimensions of the resonator, or to fluctuations of the dielectric and magnetic constants of the material in the resonator's interior.

The basic causes of eigenfrequency drift in superconducting cavity resonators were considered in section 6. When the cavity is filled with a dielectric there are important effects that can change the character of the frequency drift in dielectric resonators. We shall now consider these effects.

In Section 6 we saw that the temperature frequency coefficient of a superconducting resonator is determined by variations of the imaginary part of its surface resistance and by the finite value of the resonator's thermal expansion coefficient at low temperatures. Using equations (6.19) and (6.22), we can derive an analytic expression for the temperature frequency coefficient due to these factors

$$\frac{1}{\omega}\frac{\partial\omega}{\partial T} \approx -C\,\frac{\omega\mu_0\delta(0)}{2\Gamma(\epsilon')}\,\frac{\Delta(0)}{kT^2}\,\exp\left(-\frac{\Delta(0)}{kT}\right) - \alpha(T)\,. \quad (7.13)$$

In a dielectric resonator, by contrast with a cavity resonator, the expansion coefficient $\alpha(T)$ is controlled by the properties of the dielectric interior. This permits the temperature frequency coefficient to be reduced by the selection of a dielectric with a small thermal expansion coefficient.

Using sapphire with its high Debye temperature $(T_D - 1047\,\mathrm{K})$, it is possible to reduce considerably the

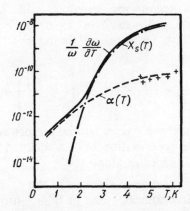

Fig. 18 Temperature dependence of temperature frequency coefficient for resonator with sapphire interior covered by Nb_3Sn film (Braginsky, Panov, and Vasiliev 1981).

contribution of the expansion coefficient to the frequency drift, since at low temperatures experimental measurements of sapphire give $\alpha(T) = (5.3 \pm 1.3) \times 10^{-13} T^3 K^{-1}$ (Vasiliev and Panov 1980). To illustrate, figure 18 shows the temperature dependence of the temperature frequency coefficient as given by equation (7.13) for a sapphire resonator covered by a film of Nb_3Sn. The curve is drawn for $\Gamma(\epsilon') = 10^2 ohm$, $\omega = 2 \times 10^{10} sec^{-1}$, and $\delta(0) = 5 \times 10^{-8} m$. The thermal expansion does not limit the value of $\omega^{-1} \partial\omega/\partial T$ down to $T \approx 2K$, where the temperature frequency coefficient is $\approx 5 \times 10^{-12} K^{-1}$.

One feature of resonators with dielectric interiors is the dependence of the resonant frequency on the dielectric constant ϵ', which changes its value when the dielectric is deformed. Small changes of the dielectric constant, and thence of the resonator frequency, can be estimated from perturbation theory (Altman 1964; Gvozdover 1956)

$$\frac{\Delta\omega}{\omega} \approx -\left(1 + \frac{1}{2}\frac{\Delta\epsilon'}{\epsilon'}\frac{1}{\beta}\right)\beta = -\left(1 + \frac{K_{\epsilon'}}{2}\right)\beta , \qquad (7.14)$$

where β is the fractional variation of the resonator dimensions and $K_{\epsilon'}$ is a coefficient characterizing the fractional changes of the dielectric constant per unit deformation of the dielectric. The value of this coefficient for sapphire can be in the range $K_{\epsilon'} \approx 1$ to 10. Taking $\beta - \alpha(T)$ in equation (7.14), one can find an expression for the temperature frequency drift including the effect of the ϵ'-variation

$$\frac{1}{\omega}\frac{\partial\omega}{\partial T} - \left[\frac{1}{\omega}\frac{\partial\omega}{\partial T}\right]_{X_i} - \left[1 + \frac{K_{\epsilon'}}{2}\right]\alpha(T). \qquad (7.15)$$

The contribution of the $K_{\epsilon'}$ term can be almost an order of magnitude larger than the contribution $\alpha(T)$ associated with the sapphire thermal expansion coefficient alone.

Analogous effects due to changes in the external pressure acting on the dielectric resonator must be taken into account. For an isotropic pressure, the strain in the resonator is

$$\beta \approx -\frac{1-2v_0}{V}P\,,$$

where v_0 is the Poisson ratio of the resonator material. In this case the "pressure frequency coefficient" is

$$\frac{1}{\omega}\frac{\partial\omega}{\partial P} \approx -\left[1 + \frac{K_{\epsilon'}}{2}\right]\left[\frac{1-2v_0}{Y}\right] \qquad (7.16)$$

and equals 10^{-9}Torr^{-1} for $K_{\epsilon'} \approx 10$. This estimate shows that the pressure of the surrounding atmosphere must be kept stable to an accuracy of $\sim 10^{-6}\text{Torr}$ in order to achieve a frequency stability of $\Delta\omega/\omega \approx 10^{-15}$.

As for a cavity resonator, in a dielectric resonator the anisotropic pressure of the oscillating electromagnetic field causes frequency drift. The drift depends on the oscillation mode, but the differences between modes are not large for dielectric resonators. Using the electromagnetic field distribution of an E_{010} mode (section 6), we can easily estimate the frequency drift due to power variations in the microwave-frequency field

$$\frac{1}{\omega}\frac{\partial\omega}{\partial W} \approx -\left[1 + \frac{K_{\epsilon'}}{2}\right]\frac{1}{2\sqrt{2}}\frac{Q_e}{\omega Y V}. \qquad (7.17)$$

For $Q_e - 10^9$, $\omega - 2 \times 10^{10}\text{sec}^{-1}$, and $V - 10^{-3}\text{m}^3$, we find that the fractional frequency drift is in the range $(6 \text{ to } 25) \times 10^{-9}\text{W}^{-1}$.

Other factors can influence the eigenfrequency of a superconducting dielectric resonator. Most of them act through variations of the imaginary part of the superconductor's surface impedance and can be estimated using the formulae given in section 6.

The most important results of filling the interiors of resonators with dielectrics are an improved reproducibility of the frequency and reduced long-term variations of resonator parameters. These features are sensitive to the quality of the adhesion of the superconducting film to its dielectric interior; the rate of diffusion of gas through the superconducting film and onto its inner working surface; and the interaction between gases dissolved in the dielectric and in the working surface of the superconductor. These effects are practical in nature and cannot be derived analytically. Experimental investigations are needed. Before showing the relevant experimental results, we will describe the technological procedures used in fabricating superconducting resonators filled with sapphire.

Among the various types of resonators, superconducting dielectric resonators oscillating in the E_{010} mode are the most convenient to fabricate and use. A capacitance coupling probe introduced along the resonator axis easily excites the appropriate electric field in resonators of this type. Depending on the length and the shape of the resonator, the geometric factor can lie in the range of $\Gamma(\epsilon') = 70$ to $100\,$ohm. For a given diameter D, the eigenfrequency in the E_{010} mode is lower than for other modes $(\lambda_e = \sqrt{\epsilon'} \times 1.305 D)$ and is independent of the length of the resonator. Figure 17 (page 68) illustrates the typical dimensions and shapes of resonators with resonant frequencies $f \approx 3\,$GHz. The interiors of these resonators are cut with diamond tools from leucosapphire which has a small concentration of paramagnetic impurities $(10^{-2}$ to $10^{-4}\%)$ and no visible gaseous occlusions. After mechanical treatment, the surface of the sapphire is carefully polished with diamond pastes until an optically smooth surface is achieved.

The eigenfrequency of the resonator is measured twice, first after cutting and then after polishing. Large deviations from the desired frequency $(\geq 100\,$MHz$)$ are reduced by additional sanding, and small errors are reduced by polishing. For preliminary measurements of the frequency, the sapphire is chemically covered by a silver film $\approx 0.5\,\mu$m thick. Such a resonator has a quality factor of $Q_e \geq 2 \times 10^3$, and the resonator frequency can be measured to an accuracy $\Delta f \approx \pm 10\,$kHz. When the desired frequency is obtained, and before the superconducting layer is applied, the sapphire is either carefully washed in a mixture of H_2SO_4 and HF and then in an ammonia solution, or it is annealed in a vacuum at a temperature of $\approx 1100°C$.

Lead and niobium films must be applied by evaporative coating techniques. Lead coating is carried out in a vacuum of $<10^{-6}$ Torr at room temperature. To get a homogeneous, tight film, the sapphire is rotated at an angular velocity of ≈ 10 rpm. To get minimal residual resistance, it is necessary to use high-purity lead ($<10^{-5}\%$ impurities). The coating rate used creates a film 0.3 to 0.5 μm thick in 10 to 15 min. This film has a light, mirror-like surface. Further increases in thickness lower the quality of the surface, which becomes loose, and mosaic cracks can appear when the resonator is cooled.

The general drawback of lead coatings is that they are short-lived. As a rule, the surface resistance of lead coatings increases from 3×10^{-7} to $(3 \text{ to } 5) \times 10^{-6}$ ohms within a week or after several cooldown cycles. As a result the resonator quality factor is strongly degraded, and the resonant frequency is shifted. However, lead coatings are suitable for short-duration measurements provided that the resonator is held at liquid-nitrogen or liquid-helium temperatures all the time.

Balalykin has developed a method for the deposition of a niobium film on a sapphire backing that involves electron-plasma evaporation in an oil-free vacuum of $<10^{-7}$ Torr (Balalykin, Zubietov, and Panov 1978). High-purity niobium that has been multiply refined by an electron beam is used. Before coating, the sapphire must be warmed up to a temperature of 800 to 900°C, and in the process of coating, the temperature of the sapphire must be held in the range of 450 to 500°C. Under these conditions, long-lived niobium films with low residual resistance can be produced.

The lowest residual resistance was obtained by Balalykin, Zubietov, and Panov when the niobium film was evaporated from two crucibles with electron guns. The distance between the crucibles was 15 cm. The resonator was suspended horizontally 12 cm above the crucible plane and rotated about its axis. The coating rate was not less than 500 Å/min., and the total thickness of the niobium film was not more than 1 to 1.5 μm. Under all these conditions the critical temperatures of the superconducting films were the same as for a massive niobium specimen, and the films were strong, long-lived, and stable, almost retaining their original parameters under storage in a dry atmosphere for one year. The resonators fabricated by this method were examined to determine their quality factors, stability, and frequency reproducibility, and to find the causes of their aging.

We will not consider here the methods by which the parameters of superconducting resonators are measured. Measurement techniques are discussed in detail by Didenko (1973) and Mende, Bondarenko, and Trubitsyn (1976). We shall only report the main experimental results for superconducting resonators with sapphire interiors (Bagdasarov, Braginsky, and Zubietov 1977; Braginsky and Panov 1979; Balalykin, Zubietov, and Panov 1978; Braginsky, Panov, and Vasiliev 1981).

Measurements of the quality factors for sapphire resonators with lead or niobium films were performed over a wide temperature range (from 10 to 1.6 K). The best quality factors were obtained when, during cooling of the resonator, the surrounding magnetic field had a strength of not more than 1.6 At/m. The best values were $Q_e = 3.5 \times 10^8$ for resonators with lead films, and $Q_e = 4 \times 10^8$ for resonators with niobium films.

Figures 19 and 20 show experimental results for the temperature dependence of the quality factor. The data are for several variants of lead and niobium films. It turned out, among other things, that a lead film applied at a rate of 3 to 5 Å sec^{-1} gives quality factors six times greater than when the film is applied at a rate of 10 to 20 Å sec^{-1} (figure 19). Nevertheless, the critical temperature for the lead film was $T_c = 7.2$ K in both cases. It was also found that the largest quality factors were the same whether the film was applied at a pressure of 10^{-5} Torr or 10^{-6} Torr.

Interesting results were obtained with niobium-coated resonators (figure 20). To find the best conditions for achieving low residual surface resistance, the films were evaporated onto sapphire backings that had different temperatures (20°C, 350°C, and 450 to 500°C). The critical temperatures for films applied under these different conditions were respectively 8.2 K, 8.6 K, and 9.2 to 9.25 K. From the curves of figure 20 it is evident that the residual resistance was lowest for films evaporated onto a 450–500°C sapphire surface. This result might be due to two factors. First, heating sapphire in a high vacuum drives foreign molecules out of its surface layers. Consequently, the first niobium layers to be applied (the ones that will interact with the microwave field most strongly) will be especially pure, thus causing a high value for the film's critical temperature (Golovashkin, Levchenko, and Motulevich 1972,

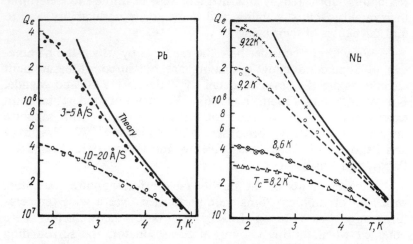

Fig. 19 Temperature dependence of Q_e for resonators with sapphire interiors covered by lead films applied at different rates by evaporative coating.

Fig. 20 Temperature dependence of Q_e for resonators with sapphire interiors and niobium films with different values of T_c.

1975). Second, the high temperature of the sapphire backing gives the applied niobium atoms an added mobility that toughens the film and thereby reduces losses.

The experimentally observed limit on the quality factor, $Q_e \approx 4 \times 10^8$ at $T \approx 2\,\mathrm{K}$, is perhaps due to residual losses in the superconducting film, because it turns out that an increase of the geometric factor Γ produces a proportional increase of the limiting quality factor. Nevertheless, dielectric and paramagnetic losses can also influence the quality factor and must be taken into account. In particular, assuming that the quality factor is determined by paramagnetic losses and allowing for the typical dependence $\chi'' \propto T^{-1}$, one can estimate the expected temperature dependence $Q_e(T)$. This dependence for a superconducting resonator with a lead coating is shown in figure 21 (dashed curve) along with experimental results for $Q_e(T)$ for one of the lead resonators.

A comparison of the curves and data in figure 21 shows that the measured value of $Q_e(T)$ is approximately 30% below the theoretical one. In addition, the theoretical curve has a maximum.

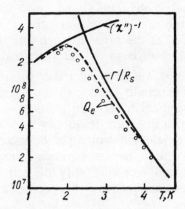

Fig. 21 Temperature dependence of Q_e for resonator with lead coating and sapphire interior with Cr^{3+} impurities. Curves are theoretically calculated; circles are experimental results.

Such a maximum was not actually observed with low-impurity sapphire interiors, but when chromium impurities were present (ruby), the quality factor was sharply limited to a peak value $Q_e \approx (1 \text{ to } 1.2) \times 10^8$ at $T \approx 2.0$ to $2.1\,K$ with a modest reduction to $Q_e \approx (0.8 \text{ to } 1.0) \times 10^8$ at $T = 1.6\,K$. This $Q_e(T)$ shows that paramagnetic chromium impurities can influence sapphire resonator quality factors.

The temperature dependence of the dielectric relaxation losses is more complicated than that of the paramagnetic (χ'') losses. Moreover, the order of magnitude of the dielectric losses is sensitive to crystal defects. According to equation (7.4), for an ideal crystal the temperature dependence of $\tan\delta$ is proportional to T^5 (for $T \ll T_D$). By contrast to the experimental data, this would lead to a rapid increase of the quality factor as $T \to 0$. But the presence of crystal defects changes sharply the temperature dependence of $\tan\delta$ and limits the resonator's microwave-frequency quality factor. With crystal defects present, analytic estimates can give only a rough value of $\tan\delta$ (see above) and, as a result, only a rough upper limit on the dielectric losses in sapphire. Using the experimental dependence $Q_e(T)$, and allowing for the finite residual resistance of the superconducting film R_0, for sapphire at $T = 2K$ and $\omega = 2 \times 10^{10} sec^{-1}$, Braginsky, Panov, and Vasiliev (1981) find tan$\delta \leq 1 \times 10^{-9}$. Recently (1984), we used the same technique and obtained tan$\delta \leqslant 2 \times 10^{-10}$ at the lower temperature $T = 1.6\,\text{K}$; this limit, at the same temperature, has been obtained by G. J. Dick and D. Strayer at Caltech (personal communication).

Until recently, it was presumed that the inadequate stability and frequency reproducibility of superconducting resonators made it impossible to use them to develop a microwave-frequency standard competitive with quantum frequency standards. But this may change due to the development of superconducting resonators with sapphire interiors, which promise significant improvements in the long-term stability and reproducibility of frequency and thereby may permit expanded applications of superconducting resonators as frequency standards.

From experimental studies of the frequency stability of various sapphire resonators we can infer the values of the resonator parameters that are needed to achieve a given level of frequency drift.

The frequency drift of sapphire resonators has been studied experimentally by Braginsky and Panov (1979), Balalykin, Zubietov, and Panov (1978), and Braginsky, Panov, and Vasiliev (1981). Figure 22 shows the fractional change in resonant frequency $\Delta\omega/\omega$ with changing temperature (solid curves) for resonators with quality factors $Q_e = 3 \times 10^8$ and $Q_e = 1.2 \times 10^8$, which had niobium films with critical temperatures $T_c = 9.25\,\text{K}$ and $T_c = 8.8\,\text{K}$, respectively. The dashed curve shows the temperature frequency coefficient $\omega^{-1}\partial\omega/\partial T$ found experimentally for the resonator quality factor $Q_e = 3 \times 10^8$ ($T_{c\,\text{Nb}} = 9.25\,\text{K}$). The experimental results are well approximated by the analytic expression

$$\frac{\omega(T) - \omega(4.2\,\text{K})}{\omega} \approx C\,\frac{\omega\mu_0\delta(0)}{2\Gamma(\epsilon')}\,\exp\left\{-\frac{\Delta(0)}{kT}\right\}$$

with the following parameters: $C = 1.9$, $\Delta(0) = 1.8kT_c$, and $\delta(0) = 4.8 \times 10^{-8}\,\text{m}$. This implies that for these resonators the frequency drift is caused mainly by the temperature dependence of the magnetic-field skin depth, and that down to $T \sim 1\,\text{K}$, the influence of the linear expansion coefficient is negligible compared to that of the surface-reactance effects of skin depth.

Fluctuations in the electromagnetic energy stored in a resonator influence the resonator's frequency. Because of its low mechanical rigidity, stored-energy fluctuations act on a cavity resonator

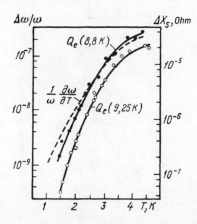

Fig. 22 Temperature dependence of temperature frequency coefficient $\omega^{-1}\partial\omega/\partial T$ (dashed curve), temperature-induced frequency changes $\Delta\omega/\omega$ (solid curves), and surface reactance $\Delta X_s = 2\Gamma\Delta\omega/\omega$ (solid curves) for resonators with sapphire interiors and niobium coatings (Braginsky, Panov, and Vasiliev 1981).

mainly through their changes of the resonator's dimensions. By contrast, dielectric resonators are sufficiently rigid mechanically that the dominant influence of energy fluctuations is through changes in skin depth.

Figure 23 illustrates the frequency drift for two sapphire-niobium resonators with different critical temperatures T_c and different critical values of the magnetic field strength H_c. The intrinsic (unloaded) quality factors of the resonators are $Q_e = 3 \times 10^8$ (for $T_c = 9.25\,\mathrm{K}$) and $Q_e = 1.2 \times 10^8$ (for $T_c = 8.8\,\mathrm{K}$). During the measurements shown, the loaded quality factor was $Q_e = 6 \times 10^7$ for both resonators, and the energy stored in each resonator was measured by several methods for greater accuracy. The measurements were performed at a temperature of $T \approx 2\,\mathrm{K}$. The temperature dependence of the observed frequencies can be explained by the temperature dependence of the skin depth, as given by the Ginzburg-Landau theory; and, practically speaking, the frequencies are not influenced by the resonator's mechanical rigidity. More specifically, the experimental results agree well with the the Ginzburg-Landau-based formula

$$\frac{1}{\omega}\frac{\partial\omega}{\partial W} \approx \frac{2\delta(0)Q_e F(\kappa)}{\Gamma(\epsilon')VH_c^2}$$

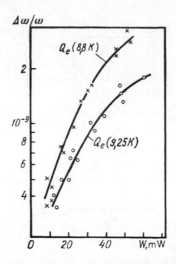

Fig. 23 Eigenfrequency as function of microwave-frequency power for resonator with sapphire interior (Braginsky, Panov, and Vasiliev 1981).

(equation [6.26]), with $\delta(0) = 4.8 \times 10^{-8}$m, $F(\kappa) = 6 \times 10^{-2}$, $H_c(9.25\,\text{K}) = 1.5 \times 10^5$At/m, $H_c(8.8\,\text{K}) = 1.2 \times 10^5$At/m, $Q_e = 6 \times 10^7$, and $\Gamma(\epsilon') = 70$ ohm. For these parameter values, the finiteness of the mechanical rigidity of the sapphire resonator contributes only 10% to the experimentally observed dependence $\omega(W)$ and thus is not a significant factor in the frequency drift due to power fluctuations. For the above parameters we find

$$\frac{1}{\omega}\frac{\partial\omega}{\partial W} \approx -3 \times 10^{-8} \text{ W}^{-1}.$$

The experimental results for the eigenfrequencies of sapphire-niobium resonators are in close agreement with the theoretical estimates discussed above for $\omega^{-1}\,\partial\omega/\partial T$ and $\omega^{-1}\,\partial\omega/\partial W$. These results allow us to infer the temperature and power stabilities required for a desired level of frequency drift. For instance, to get a fractional frequency stability of $\Delta\omega/\omega \approx 10^{-14}$ at $T = 1.6\,\text{K}$, it is necessary to hold the temperature of the resonator's walls constant to within an accuracy of $\approx 3 \times 10^{-6}$K, and to hold the power constant to within 3×10^{-7}W.

In applications of superconducting resonators it is important to have good frequency reproducibility and good long-term frequency stability. Studies of cavity resonators have shown that their frequency reproducibility after warmups and cooldowns is 10^{-7} at best, and that their long-term stability is strongly dependent on storage conditions. As a rule, the quality factors of cavity resonators decrease by several orders of magnitude after a single interaction with the atmosphere. Therefore, in order to hold their parameters constant, one must always keep cavity resonators in a high vacuum and at a low temperature.

By contrast, superconducting resonators with sapphire interiors exhibit good frequency reproducibility and good long-term stability, and they remain in good condition even during storage at ordinary room temperature. More specifically, the experimentally observed fractional frequency changes of niobium-sapphire resonators did not exceed 10^{-9} after many cycles (>10) of cooling and heating. The quality factors of these resonators ($Q_e = (2.5 \text{ to } 3) \times 10^8$) were unchanged by ordinary storage conditions at room temperature for a year, and the constancy of the frequencies during this time was at least as good as $\Delta\omega/\omega \lesssim 3 \times 10^{-9}$, although the resonators were cooled and heated about 30 times.

8. High-Quality Dielectric Ring Resonators

It is possible to produce high-quality electromagnetic ring resonators without metal coatings because of the phenomenon of total internal reflection at the boundary between two dielectrics. By analogy with acoustic resonators, sometimes one describes such a resonator as a "whispering-gallery resonator." Dielectric ring resonators (closed dielectric waveguides) have recently been used extensively in microwave-frequency techniques. At room temperature their quality factors can reach several tens of thousands (Dobrosmyslov and Vzyatyshev 1973; Vzyatyshev and Dobrosmyslov 1977).

Because of the absence of skin effects and of temperature dependences of surface resistance, such resonators hold promise for achieving small values of $\omega^{-1} \, \partial\omega/\partial T$ at low temperatures. Therefore, it is possible that these resonators will be convenient as stable filters and as reference elements for secondary frequency standards. At liquid-helium temperatures, the value of $\omega^{-1} \, \partial\omega/\partial T$ for these resonators is close to the linear expansion coefficient $\alpha(T)$. For sapphire, $\alpha(T) \approx 5 \times 10^{-12} \mathrm{K}^{-1}$ at $T = 2\mathrm{K}$. By contrast, for a niobium resonator, $\omega^{-1}\partial\omega/\partial T \approx 4 \times 10^{-9}\text{K}^{-1}$ even at $T = 2\text{K}$ due to the temperature dependence of the surface resistance (see section 6 for details). Furthermore, the small value of $\tan\delta \approx 10^{-9}$ observed for sapphire at low temperatures and the absence of the kinds of dissipation typical of metal resonators suggest that it may be possible to achieve large quality factors for these resonators.

By contrast with closed metal resonators, ring resonators produce electromagnetic radiation that causes radiative losses. We consider now the various factors that give rise to radiative losses (Braginsky and Vyatchanin 1980):

Radiative losses due to curvature of the waveguide. Outside the ring of the waveguide, the phase velocity of the electromagnetic wave increases with distance, until it becomes equal to c. Because of this behavior, part of the electromagnetic field is "stripped off" and becomes outflowing radiation at large distances. A numerical analysis, performed for $\epsilon' = 8.5$ and for E_{n1} and H_{n1} oscillation modes, indicates that this effect need not prevent the achievement of large quality factors. For instance, if $n = 20$ ($R/\lambda = 1.4$ where R is the radius of the the resonator's outer wall and λ is the wavelength in vacuum), the computed quality factor is $Q_e \gtrsim 6 \times 10^{11}$; and even if $n = 16$ ($R/\lambda = 1.15$), then $Q_e \gtrsim 10^9$.

Radiative losses due to spatial variations of the dielectric constant inside the resonator. Variations of the dielectric constant ϵ' are caused primarily by the anisotropy of sapphire. If the axis of a ring resonator coincides precisely with the crystal's principal axis, these variations vanish. The crystal blocks in a real crystal can be characterized by the mean angle $\Delta\varphi$ of variations in the crystal's principal axis and the mean size b of the crystal blocks. For small values of $\Delta\varphi$, the variations in ϵ' from one block to another for directions orthogonal to the axis are $\Delta\epsilon' \approx (\epsilon'_\parallel - \epsilon'_\perp)(\Delta\varphi)^2/2$.

Radiative losses due to such ϵ' variations can be estimated by using the following expression for the Q_e of a planar waveguide (low oscillation modes; Marcuse 1972; Shevchenko 1969):

$$(Q_e)_{\epsilon'} \gtrsim \frac{2\pi^2 d(\epsilon')^3 \sqrt{\epsilon'(\epsilon'-1)}}{\lambda(\Delta\epsilon'_\perp)^2} \times \frac{b}{\lambda}, \qquad (8.1)$$

where d is the halfwidth of the waveguide. This expression is correct if $V = (2\pi d/\lambda)\sqrt{\epsilon'-1} > \pi$ and $\epsilon' \gg 1$. For $\Delta\varphi = 1°$, $V = 5$, and $\epsilon' = 8.5$, the quality factor is $(Q_e)_{\epsilon'} \gtrsim 5 \times 10^{10} b/\lambda$.

Radiative losses due to roughness or to smooth inhomo-geneities of the resonator's geometry. We can estimate these losses by using formulas which have been derived for optical fibers (Marcuse, 1972; Shevchenko, 1969). An upper limit on $(Q_e)_d$ for a planar waveguide can be found from the expression

$$(Q_e)_d \gtrsim \frac{d^2}{(\Delta d)^2} \frac{1}{V} \left[\frac{BV}{d} + \frac{d}{BV} \right] \frac{A}{\cos^2 kd} , \qquad (8.2)$$

where Δd is the rms roughness (inhomogeneity), B is the correlation length of the roughness (inhomogeneity), k is the transverse wave vector, and the dimensionless factor A lies in the range $1 < A < \epsilon'$. If $B = 10d$, $d = 1\,cm$, $\cos^2 kd \approx 0.1$, and $A \approx \epsilon'$, then $(Q_e)_d \approx 10^{10}$ for $\Delta d \approx 3 \times 10^{-4}cm$.

These estimates show that radiative losses in dielectric ring resonators need not prevent one from achieving quality factors close to $Q_e \approx (\tan\delta)^{-1} \approx 10^9$ at liquid-helium temperatures. Because the effort to develop cryogenically cooled ring resonators is relatively new, there exist only preliminary quality-factor measurements for them (Braginsky, Panov, and Timashev, 1982). The resonators used in these measurements were made from leucosapphire grown by the method of oriented crystallization, possessing a concentration of dislocations of $\approx 10^3 cm^{-2}$ and a block structure with an inhomogeneity angle $\Delta\varphi \approx 1°$. The resonator rings had an outer diameter of 10 cm, an inner diameter of 6 cm, and a ring thickness of 1.8 cm. The surface of each resonator was polished to optical flatness. Inhomogeneous variations in the resonator dimensions were no more than $10^{-3}cm$, and the difference between the normal to the ring plane and the principal crystal axis was no larger than 1°. The resonator was secured between two thin-walled polystyrene rings. Oscillations in the 8 to 10 GHz band were driven by a capacitive probe which was placed at a distance of 2 to 10 mm from the resonator's surface. The probe was connected to a coaxial transmission line that was driven by a microwave oscillator with variable frequency, and the resonator's response was monitored by a quality factor measuring system.

The highest quality factors achieved in these measurements at $T = 300\,K$, $T = 77\,K$, and $T = 8\,K$, respectively, are

$$Q_e = 2.5 \times 10^5 , \qquad Q_e = 5.7 \times 10^7 , \qquad \text{and} \quad Q_e = 6 \times 10^8 .$$

For different oscillation modes with frequencies differing by no more than a few megahertz, the quality factors differed by as much as two or three orders of magnitude. To exclude the effects of losses due to surface impurities at these levels of Q, it is necessary to spend many hours cleaning the resonator in heated, aggressive chemical liquids, and then wash it in distilled water, and place it in an oil-free high-vacuum cryostat. This type of resonator permits one to obtain information on the dependence of $\tan\delta$ over a wide range of temperatures. Figure 24 shows the dependence of Q^{-1} on temperature from 300 to 3.5 K, which was obtained with one such disk-type resonator at a frequency of $\simeq 10\,\text{GHz}$ (Braginsky *et al.* 1985). Notice that the observed behavior, $Q^{-1} \propto T^{5 \pm 0.3}$ from $T = 250$ to $60\,\text{K}$ is in good agreement with the Gurevich formula (7.4).

The value $Q_e = 6 \times 10^8$ is obviously not a fundamental limit but rather was restricted by imperfect resonator fabrication or by impurities in the monocrystals. It should be noted that some delicate surface effects could be measured with these resonators because a single monomolecular layer on the surface produces an eigenfrequency shift for a given mode that is rather large ($\Delta\omega/\omega \approx 10^{-9}$).

D. Blair and S. Jones (1985) have recently proposed that high Q values might be achieved by combining the idea of a ring resonator with a superconductive resonator. A whispering-gallery mode is excited in a sapphire ring or disk as just described, and a superconducting screen is placed around the resonator to elimate radiative losses. In one of the first versions of this scheme, Blair and Jones (1985) have obtained $Q \simeq 1.2 \times 10^9$ at 1.5 K in the 10 GHz region.

High-quality dielectric ring resonators can also be developed for optical frequencies. The loss factor of 0.2 db/km achieved for quartz fibers (Arecchi and Schulz-DuBois, eds. 1972) corresponds to a quality factor of $Q_e \approx 10^{11}$ for an optical ring resonator made from such a fiber. Unfortunately, present methods of fiber welding produce large radiative losses. If this technical limitation is overcome, optical fiber techniques may produce systems with quality factors exceeding that of a 100 m Fabry-Perot resonator.

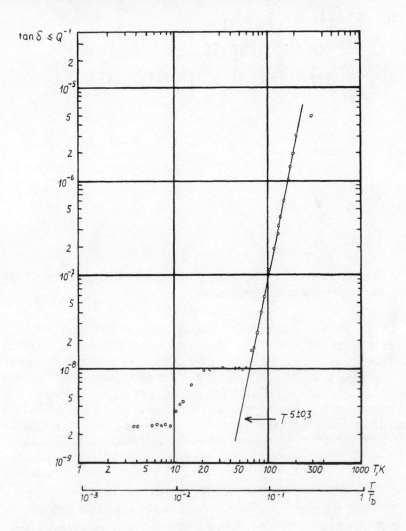

Fig. 24 Temperature dependence of Q^{-1} for 10 GHz whispering gallery electromagnetic mode in disk-shaped sapphire-crystal ring resonator (Braginsky *et al.* 1985).

IV   High-Quality
Electromagnetic Resonators
in Physical Experiments

9. Electromagnetic Self-excited Oscillators Stabilized by High-quality Superconducting Resonators

In physical experiments, the accuracy of absolute measurements is generally lower than the accuracy of differential measurements (which are the foundation for most of experimental physics). For a long time, this was not true of the stability of self-excited electromagnetic oscillators. For many years the hydrogen frequency standard (hydrogen maser) had both the greatest absolute accuracy and the greatest long-term (> 100 sec) frequency stability of all oscillators. Moreover, the absolute errors of the hydrogen-maser frequency standard were less than the short-term instabilities of almost all the secondary frequency standards that are generally used for differential measurements.

This unusual situation has begun to change as a result of the development of self-excited electromagnetic oscillators stabilized by high-quality superconducting resonators. M. S. Khaikin (1961) was the first to propose the idea of achieving high frequency stability this way. Since then, many types of self-excited oscillators stabilized by high-quality resonators have been developed and investigated, and when considerable progress was made in the fabrication of high-quality resonators, the resonator frequency stability was correspondingly improved. The short-term ($\tau = 10^{-3}$ to 10^{+4} sec) frequency stability of superconducting resonators at the time of this writing is better than the stability of the hydrogen-maser standards. Nevertheless the "competition" between hydrogen frequency standards and secondary frequency standards with superconducting resonators continues. Work toward the improvement of hydrogen standards (Vessot, Levine, and Mattison 1977) promises a level of short-term (10^{-3} to 10^{+4} sec) stability in the range $\Delta\omega/\omega \approx 10^{-17}$ or less in the near future. Such a level of stability has not been achieved for superconducting resonators. For a description of the present state of the art in the frequency stabilization of some types of electromagnetic resonators, see Turneaure *et al.* (1983).

In this section we consider the basic methods used to obtain small fractional frequency variations ($\Delta\omega/\omega$) with high-quality resonators. Before describing advances in this field, we discuss how the quality factor influences the frequency stability of electromagnetic self-excited oscillators and what might be the limits on their stability.

In quantum electronics the following well-known formula describes the limit on the frequency stability of a self-excited oscillator due to vacuum fluctuations of the electromagnetic field (Townes-Schawlow formula)

$$\frac{\Delta\omega}{\omega} \approx \left[\frac{\hbar\omega}{4WQ_e^2\hat{\tau}} \right]^{\frac{1}{2}}. \tag{9.1}$$

Here W is the electromagnetic power in the oscillator, Q_e is the quality factor of the electromagnetic resonator, ω is the frequency of oscillation, and $\hat{\tau}$ is the averaging time. This limit was derived under the assumption (see, for example, Lamb 1965) that the resonator temperature is low enough that $kT_e < \hbar\omega/2$, and the eigenfrequency ω_r of the resonator is fixed.

According to equation (9.1), an unlimited increase in W or Q_e will result in values of $\Delta\omega/\omega$ that are as small as desired. It is clear that the constancy of the resonator's eigenfrequency ω_r is an unphysical assumption, and that it is essential to take into account the many possible causes of fluctuations in ω_r. This goal was the starting point for the work of Braginsky and Vyatchanin (1978), Braginsky, Vyatchanin, and Panov (1979), and Panov and Rudenko (1979), which estimated the influence of a variety of fundamental phenomena (phenomena that cannot be removed even in principle) on the value of ω_r and thence on the limiting frequency stability of a self-excited electromagnetic oscillator. We will briefly discuss these effects.

Thermodynamic temperature fluctuations together with the finite linear expansion coefficient cause fluctuations of ω_r given by

$$\left(\frac{\Delta\omega}{\omega}\right)_r \approx \frac{2\gamma}{3YV}\left[\frac{kT_e W(\tau_T^*)^2}{\hat{\tau}}\right]^{1/2}. \tag{9.2}$$

Here Y is the Young's modulus of the resonator material, V is the volume of the resonator (the resonator is assumed to be solid), τ_T^* is the thermal relaxation time, and γ is the Grüneisen constant. Lowering T_e to $kT_e < \hbar\omega_r$ eliminates thermodynamic temperature fluctuations, but the absorption of a single quantum in the resonator (due to vacuum fluctuations) creates a shot effect: the temperature jumps by $\delta T_e = \hbar\omega/C_V V$, causing fluctuations of ω_r given by

$$\left(\frac{\Delta\omega}{\omega}\right)_r \approx \frac{\gamma}{3YV}\left[\frac{2\hbar\omega W(\tau_T^*)^2}{\hat{\tau}}\right]^{1/2}. \tag{9.3}$$

It follows from this formula that the only way to reduce $(\Delta\omega/\omega)_r$ for a given power W is to decrease τ_T^*; i.e., to improve the thermal contact between the thermostat and the resonator.

When τ_T^* is small enough, a final fundamental effect must be taken into account, namely, the ponderomotive shot noise of the pressure on the resonator walls. The creation or annihilation of a single electromagnetic quantum $\hbar\omega$ in the resonator creates a jump in the pressure on the resonator walls, which in turn causes a jump in the resonator's eigenfrequency. If many quanta are created and annihilated in the resonator during the averaging time $\hat{\tau}$, then

$$\left(\frac{\Delta\omega}{\omega}\right)_r \approx \frac{1}{4YV}\left[\frac{2W\hbar\omega(\tau_e^*)^2}{\hat{\tau}}\right]^{1/2}, \tag{9.4}$$

where $\tau_e^* = 2Q_e/\omega$ is the relaxation time for electromagnetic oscillations.

By comparing equations (9.4) and (9.3), we see that ponderomotive shot noise is dominant if $\tau_T^* < \tau_e^*$ (because $\gamma \approx 1$ for typical solids). Equation (9.4) also shows that, by contrast with the Townes-Shawlow limit (9.1), a decrease of W is accompanied by a decrease of $(\Delta\omega/\omega)_r$. Combining equations (9.1) and (9.4), we obtain expressions for the best frequency stability achievable for self-excited electromagnetic oscillators with real resonators

$$\left(\frac{\Delta\omega}{\omega}\right)_{min} \approx \left(\frac{\hbar}{\sqrt{2}YV\hat{\tau}}\right)^{1/2}, \tag{9.5}$$

and for the optimal electromagnetic power corresponding to the best stability

$$W_{\text{opt}} \approx \frac{YV\omega}{\sqrt{2}Q_e^2} \, . \tag{9.6}$$

For $Y = 4 \times 10^{11}\text{Pa}$, $V = 10^{-5}\text{m}^3$, $\hat{\tau} = 10\,\text{sec}$, $\omega = 2 \times 10^{10}\text{sec}^{-1}$, and $Q_e = 10^{10}$, equations (9.5) and (9.6) give

$$(\Delta\omega/\omega)_{\text{min}} \approx 1 \times 10^{-21} \, , \quad \text{and} \quad W_{\text{opt}} \approx 1 \times 10^{-3}\text{W} \, .$$

As we shall see below, this value of $(\Delta\omega/\omega)_{\text{min}}$ is approximately five orders of magnitude less than the values that have actually been attained. In other words, there is plenty of room for improving the most fundamental characteristic of an electromagnetic self-excited oscillator, namely, its frequency stability.

Equation (9.5) can be derived alternatively from more fundamental considerations by taking account of the fact that the frequency shift $\Delta\omega/\omega$ gives an observer information on changes $\overline{\Delta l}$ of the linear dimensions l of the resonator, averaged over the time $\hat{\tau}$: $\Delta\omega/\omega \approx \overline{\Delta l}/l$. Now, for any continuous measurement of the coordinate of a mechanical oscillator, the minimum error allowed by the Heinsenberg uncertainty principle is

$$(\overline{\Delta l})_{\text{min}} \approx \left| \frac{\hbar}{M\omega_M} \right|^{\frac{1}{2}} (\omega_M\hat{\tau})^{-\frac{1}{2}} \, .$$

Here M and ω_M are the mass and angular frequency of the mechanical oscillator. Allowing for only the fundamental normal mode of mechanical vibration of our electromagnetic resonator, and expressing M and ω_M in terms of Y and V, we easily derive

$$\left| \frac{\Delta\omega}{\omega} \right|_{\text{min}} \approx \frac{\overline{\Delta l}}{l} \approx \left(\frac{\hbar}{\sqrt{2}YV\hat{\tau}} \right)^{\frac{1}{2}} \, ,$$

which is equation (9.5).

Our derivations of equations (9.5) and (9.6) did not rely on any assumption that the high-quality resonator is fabricated from a superconductor. The skin depth δ in a superconductor depends on the strength of the microwave-frequency magnetic field (Ginzburg and Landau 1950)

$$\delta(H) \approx \delta(0) \left[1 + \frac{\kappa}{6\sqrt{2}} \frac{H^2}{H_c^2} \right] , \tag{9.7}$$

where H_c is the strength of the critical field and κ is the Ginzburg-Landau parameter. This effect leads to an additional dependence of ω_r on the electromagnetic power W, since $W \propto H^2$. This result is completely analogous to the ponderomotive effect considered above [equation (9.4)]. Both cause fluctuations in the effective dimensions of the resonator and thence in the resonator frequency; these fluctuations are proportional to $W^{1/2}$ when averaged over time. Thus, from the dependence of skin depth on field strength [equation (9.7)], one can derive stability limits analogous to those of equations (9.5) and (9.6). However, these stability limits will be less constraining than the ponderomotive effect when H_c is large and $\delta(0)$ is small, more specifically when

$$\frac{H_c^2}{\delta(0)} \gtrsim \frac{8\kappa Y}{l} , \tag{9.8}$$

where l is the linear size of the oscillator.

Thus, if a superconducting resonator is fabricated from a sufficiently "good" superconductor, the limiting stability is given by equation (9.5). We have omitted other details concerning the conditions of applicability of equations (9.5) and (9.6). A more complete analysis can be found in Braginsky and Vyatchanin (1978); Braginsky, Vyatchanin, and Panov (1979); and Panov and Rudenko (1979). In the last of these references the "classical" regime $kT_e \gg \hbar\omega$ is also discussed. Classical fluctuations of the resonator's dimensions limit stability in this case, and it is necessary to take into account such detailed classical effects on electromagnetic self-excited oscillators as the stiffness of the limit cycle. The analysis of Panov and Rudenko gives details of the resulting influence of classical ω_r fluctuations on the spectrum of the output frequency fluctuations of a self-excited oscillator.

The resonator's quality factor Q_e does not appear explicitly in the fundamental limit (9.5) for the stability of a self-excited oscillator. However, it is clear that a small value of $\Delta\omega/\omega$ is only possible in practice when the quality factor is high. In particular, equation (9.5) is valid only for $kT_e < \hbar\omega/2$, which implies that T_e must be close to or less than liquid-helium temperature. Moreover, to achieve the minimum value (9.5) of $(\Delta\omega/\omega)_{min}$, the resonator must contain the power W_{opt} given by equation (9.6). Now, some of this power is unavoidably dissipated in the resonator and heats it. The estimate given above ($W_{opt} \approx 10^{-3}W$ for $Q_e = 10^{10}$) is reasonable in terms of refrigeration technology if $T_e \approx 0.3$ to $1\,K$. But if the experimenter has a resonator with a quality factor of, say, 10^7, then for the same Y, V, and ω as above, the power required for maximum stability is $W_{opt} = 10^3$W. Even at liquid-helium temperature, the heat produced by such power is too great for removal from the resonator without degrading its performance.

Thus, excellent frequency stability can only be achieved with an electromagnetic resonator possessing a sufficiently large quality factor.

The most successful experiments thus far in stabilizing the frequencies of electromagnetic self-excited oscillators have been those performed at Stanford University in the 1970s by Stein and Turneaure (1973, 1976, 1978) and Stein (1974, 1975). Using an automatic frequency control (AFC) circuit, Stein and Turneaure obtained record stabilities for microwave-frequency self-excited oscillators with stabilizing superconducting cavity resonators ("superconducting cavity-stabilized oscillators" or "SCSOs"). The fractional frequency drift was no more than 3×10^{-14}/day, and the maximum frequency stability was $\Delta\omega/\omega \approx 3 \times 10^{-16}$ for measurement times of 3×10^2 to 3×10^3 sec.

The development of electromagnetic self-excited oscillators with high frequency stability is important for many precise physical experiments. For this reason we shall describe the methods used by Stein and Turneaure in the construction of their SCSOs.

Stein and Turneaure used a modified version of the AFC system developed by R. V. Pound (1946). A simplified diagram of the system is shown in figure 25. The system was used to control the frequency of a Gunn diode-powered oscillator ("generator") with an output $W_{out} \approx 50$mW at 8.6 GHz. An E_{010}-mode niobium cavity resonator with a quality factor of $Q_e \approx 10^9$ to 10^{10} was used as the stabilizing element. The resonator was placed in a vacuum cryostat at temperature $T = 1.3$K, stable to $\pm 10^{-6}$K. The temperature stabilization system provided a constant resonator temperature to an accuracy of $\Delta T \approx 10^{-5}$K over several days. The Gunn-diode generator and the control circuit were at room temperature. This allowed the use in the servosystem of sophisticated equipment that would not have functioned at low temperatures.

The frequency stabilization was accomplished by a comparison of the eigenfrequencies of the niobium resonator and the Gunn-diode generator. Part of the Gunn-diode generator power was modulated at a frequency of 1 MHz (instead of the usual 30 MHz) and used to drive the niobium resonator, producing an error

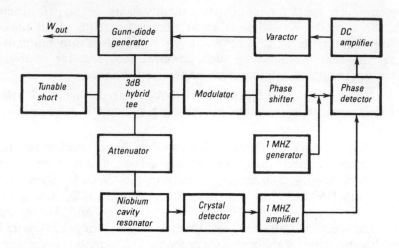

Fig. 25 Diagram of self-excited oscillator stabilized by automatic frequency control system based on superconducting cavity resonator ("SCSO") (Stein and Turneaure 1973; Pound 1946).

signal proportional to the mismatch between the Gunn diode's frequency and the center of the niobium resonator's resonance curve. The output of the driven resonator was converted into an amplitude-modulated signal in which the degree of modulation was proportional to the diode-resonator frequency difference. By synchronous detection, the envelope of the output signal was converted into an error signal which, after some amplification and filtering, was applied to the fine-adjustment element of the Gunn-diode generator, thereby correcting its frequency.

In order to achieve their remarkable frequency stability, Stein and Turneaure had to solve complicated practical problems in the construction and environment of their SCSO. For example, deformation of the resonator by the earth's gravitational forces changed its eigenfrequency by $\Delta\omega/\omega \approx 10^{-9}$; and because of the gradient of the Earth's gravitational field, a change of 1 cm in the distance from the Earth's center to the resonator shifted the eigenfrequency by $\Delta\omega/\omega \approx 10^{-16}$. Such distance changes, caused by tides in the solid earth, have amplitudes ≥ 30 cm and produced correspondingly large diurnal frequency variations.

In addition, changes in the tilt and angular orientation of the cryostat could produce significant frequency changes. In these experiments, the fractional frequency change was equal to 3×10^{-14}/arcsec. In order to reduce the influence of these effects, a servosystem was developed to hold fixed the orientations of the cryostat and all other parts of the system.

The frequency stability was also influenced by changes in the resonator's dimensions due to the motion of dislocations, variations of rigidity at places where the resonator was attached to the transmitting waveguide, seismic and acoustic disturbances, and penetrating radiation. The AFC circuit restricted the frequency stability as a result of thermal drifts of the phase-modulation detector and drifts of the DC amplifiers in the frequency control circuits of the Gunn-diode generator.

The fractional frequency stability over short periods of time was measured by comparing the frequencies of three identical SCSOs, and the long-term frequency drift was measured relative to a cesium frequency standard. The fluctuations of frequency with time were characterized by the "Allan (1966) variance," according to which the most probable frequency change during a time $\hat{\tau}$ is

$$\Delta\omega(N, T, \hat{\tau}) = \left\langle \frac{1}{N-1} \sum_{n=1}^{N} \left| \overline{\omega}_n - \frac{1}{N} \sum_{k=1}^{n} \overline{\omega}_k \right|^2 \right\rangle^{1/2} . \qquad (9.9)$$

Here T is the averaging time for individual frequency measurements, N is the number of measurements made (each separated from the next by time $\hat{\tau}$), and $\overline{\omega}_n$ and $\overline{\omega}_k$ are the measured frequency values. If $N = 2$, then equation (9.9) has the simple form:

$$\sigma_\tau = \Delta\omega(2, T, \hat{\tau}) = \left\langle \frac{(\overline{\omega}_{k+1} - \overline{\omega}_k)^2}{2} \right\rangle^{1/2} . \qquad (9.10)$$

This Allan variance can be used to characterize the frequency stability during a given time $\hat{\tau}$.

Figure 26 shows the time dependence of the Allan variance [fractional frequency stability; equation (9.10)] for the Stein-Turneaure SCSOs. There are three different time ranges: For short times, $\hat{\tau} < 30\,\text{sec}$, the stability curve is described by $\Delta\omega/\omega \approx 10^{-14}\hat{\tau}^{-1}$, which is similar in time dependence to the $\Delta\omega/\omega$ that would be produced by phase shifts due to natural electromagnetic fluctuations. The second region, $30\,\text{sec} < \hat{\tau} < 100\,\text{sec}$, exhibits the minimum frequency drift, $\Delta\omega/\omega \approx 3 \times 10^{-16}$. The third region, $\hat{\tau} > 100\,\text{sec}$, is characterized by a

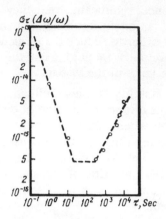

Fig. 26 Allan variance for fractional frequency fluctuations of SCSO (Stein and Turneaure 1973, 1976; Stein 1974).

deterioration of stability with increasing $\hat{\tau}$, which appeared to be due both to changes in the individual elements of the AFC electronic circuit and to frequency drifts of the superconducting resonator.

The long-term frequency drifts of the Stein-Turneaure SCSOs were measured by comparing their frequencies with cesium frequency standards. To avoid errors in the magnitudes and the signs of the frequency drifts, several cesium standards were used, and the measurements were performed over a period of twenty days. During these measurements the mean value of the linear fractional frequency drifts of the SCSOs was no more than 3×10^{-14}, and the drift signs were different for different measurement runs. Along with the linear drift, diurnal periodic variations of frequency were observed. According to Stein and Turneaure (1973, 1976) and Stein (1974) these variations were produced by tidal effects and by slow displacements of the floor and walls of the laboratory that housed the oscillators.

In addition to measurements in the time domain, Stein and Turneaure measured the spectral density $S_\varphi(\omega - \Omega)$ of the SCSO phase fluctuations φ as a function of the angular frequency $\omega - \Omega$ of detuning from the carrier frequency Ω. The spectral density curve is shown in figure 27. The spectral density can be divided into several frequency ranges: In the central band from

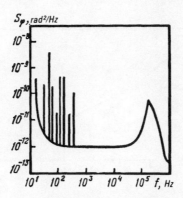

Fig. 27 Spectral density of phase fluctuations for a SCSO (Stein and Turneaure 1973, 1976, 1978; Stein 1974).

$f \equiv \omega - \Omega/2\pi = 10^2$ to 5×10^4Hz, the spectral density, being equal to a constant 10^{-12}rad^2/Hz, corresponds to white noise. This minimum value of the spectral density S_φ must have been due to noise in the amplitude detector of the AFC circuit, because electromagnetic fluctuations in the resonator itself would have given a lower limit on S_φ.

In the range $\omega - \Omega/2\pi < 100$Hz, the spectral density of the phase fluctuations has the form $S_\varphi(\omega - \Omega) \propto (\omega - \Omega)^{-3}$, which reveals the dominant importance of flicker noise. The peak that appears near 250 kHz is associated with the frequency response of the servosystem, which has an upper frequency limit of $\approx$300 kHz. In this region phase fluctuations of the Gunn-diode generator are not eliminated by the servosystem; consequently, S_φ shows an increase. The sharp peaks in the 30 to 500 Hz region are caused by mechanical degrees of freedom in the resonance circuit of the SCSO, which are driven by seismic vibrations, acoustic disturbances, and natural fluctuations in the mechanical elements of the SCSO.

To recapitulate, the most important achievement of Stein and Turneaure was their successful construction of SCSOs, which set a record $\Delta\omega/\omega \approx 3 \times 10^{-16}$ for short-term fractional frequency stability and also had excellent long-term stability, $\Delta\omega/\omega \approx 3 \times 10^{-14}$/day.

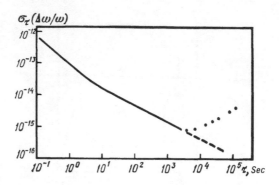

Fig. 28 Allan variance for fractional frequency fluctuations of a hydrogen-maser frequency standard (Vessot, Levine, and Mattison 1977).

These SCSOs were developed using superconducting resonators without any reliance on self-sustained maser oscillations. The SCSOs had a reasonable power output and their operating frequencies were determined solely by the geometric dimensions of the resonator. Unfortunately, the frequency reproducibility of these SCSOs was not good. It is enough to say that any warmup and cooldown of the superconducting cavity caused a fractional frequency shift greater than 10^{-9}. On the other hand, the frequency stability achieved by these SCSOs was better than the best quantum standards for times shorter than 3×10^3 sec.

To illustrate this fact, figure 28 shows a plot of the fractional frequency stability for the world's best hydrogen masers, those developed at the Harvard/Smithsonian Center for Astrophysics by Vessot, Levine, and Mattison (1977). A comparison of data for the two types of oscillators (Figures 26 and 28) shows that the SCSO had the better stability for time intervals up to 3×10^4 sec, though for longer intervals the superiority of masers is obvious.

SCSOs are too complex and delicate for practical laboratory applications. However, one should not conclude that a frequency stability level of $\Delta\omega/\omega \approx 10^{-16}$ is approachable only through such sophisticated techniques. For instance, if a Nb_3Sn resonator with a sapphire interior (similar to the resonator described in Section 7) instead of a niobium cavity resonator were used, then because of the sapphire's greater mechanical rigidity, the influence of mechanical vibrations and tilts would be much weaker. Because of the higher critical temperature of Nb$_3$Sn, either the constraints on temperature stability would not be so severe or a smaller value of $\Delta\omega/\omega$ could result from the same temperature stability. It is also safe to assume that the "solid-state" alternative would give a much better frequency reproducibility. In other words, we feel that the record results achieved in the Stein-Turneaure experiments will either be surpassed in the not-too-distant future or will be obtained much more easily.

G. J. Dick and D. M. Strayer (1984) have proposed using a maser-type pump for SCSOs and also for an all-cryogenic, three-resonator cavity with a superconductive film. It is very likely that these ideas will help to overcome the problems that appear due to high AC energy flux if tunnel or Gunn diodes are used.

A frequency stability level of $\Delta\omega/\omega \approx 10^{-15}$ to 10^{-16} is not always needed for practical laboratory applications. When a frequency stability of $\Delta\omega/\omega \approx 10^{-11}$ to 10^{-12} is sufficient, one can use high-quality electromagnetic resonators in much simpler ways than those developed by Stein and Turneaure (Mende, Bondarenko, and Trubitsyn 1976; Braginsky and Panov 1979; Braginsky, Panov, and Vasiliev 1981; Minakova *et al.* 1978; Panov 1980).

10. Applications of Superconducting Resonators in Radiophysical Measurements

The high quality factors Q_e and stable eigenfrequencies f_r of superconducting resonators at liquid-helium temperatures make them useful as sensing devices in various radiophysical measurements. Several typical examples of such applications are described in this section.

Measurements of very small mechanical vibrations. Superconducting resonators make possible the monitoring of very small mechanical vibrations with amplitudes less than 1×10^{-16} cm in a fairly simple way. The apparatus used, called a "parametric, capacitive transducer," consists of a capacitor whose capacitance varies as the distance d between its plates changes. It is this distance that one seeks to measure. The capacitor together with an inductor forms a resonator which either is put into a self-excited circuit or is excited at a "pump" frequency f_p, near its resonant frequency, by an external oscillator. In the first case, changes of the distance d

between the "plates" cause changes in the self-excited oscillator's resonant frequency; in the second case, the amplitude or phase of oscillation in the transducer's resonant circuit is changed. One or the other type of change is registered in subsequent circuits.

Figure 29 shows a test set up for such a parametric capacitive transducer (Maxwell 1964; Braginsky, Panov, and Popel'nyuk 1981). The capacitor whose plate distance d was to be monitored and the associated inductor were combined in a single electromagnetic cavity resonator (1 in figure 29). The resonator, 30 mm in diameter and 4 mm in depth with a central, reentrant stub, was fabricated from niobium. The resonator's reentrant stub and its top face ("locking diaphragm") made up the capacitor, and the distance d between them (the "gap") was the quantity to be monitored. The resonator had an electromagnetic eigenfrequency $f_r \approx 3\,\text{GHz}$, a capacitance $C_r - 1.14\,\text{pF}$, and a capacitance gap $d - 3 \times 10^{-6}\,\text{m}$. At a temperature $T - 4.2\,\text{K}$, the quality factor of the loaded resonator was $Q_e - 4 \times 10^4$. The coupling between the resonator and the external circuitry was achieved with two movable inductance loops, one used to drive the resonator from a pump oscillator with pump frequency f_p, and the other to extract the resonator's response signal. Thus, the resonator was a series element in the measuring circuit.

In the setup shown in figure 29, mechanical vibrations of the locking diaphragm labeled 3 at a frequency f_m produced oscillations of the capacitance gap d that modulated the resonator's eigenfrequency, generating a signal at frequencies $f_p \pm f_m$ in the resonator's output. The thickness of the locking diaphragm was 10 mm and its diameter was 30 mm. The rigidity (spring constant) of the diaphragm to a force applied to its center was 3.1×10^9 N/m. The mechanical eigenfrequency of the diaphragm was $\simeq 40\,\text{kHz}$. Vibrations of the diaphragm were driven and their amplitude was calibrated using the electrostatic force in a special calibrating capacitor. One of the plates of the capacitor was placed on a thick niobium disk that was fixed above the diaphragm. The central part of the diaphragm served as the second capacitor plate. The gap of this calibration capacitor was $d_{\text{cal}} - 3.4 \times 10^{-4}\,\text{m}$ and its capacitance was $C_{\text{cal}} - 1.25\,\text{pF}$. The sum of a constant voltage U_c and an alternating voltage $U_\sim - U_0\sin\omega_m t$ ($U_c \gg U_0$) was applied to the calibrating capacitor. The resulting Coulomb force caused vibrations in the diaphragm of an amplitude reliably computed to be $x - 1.25 \times 10^{-18} U_c U_0\,\text{m}$ (where U_c and U_0 are in

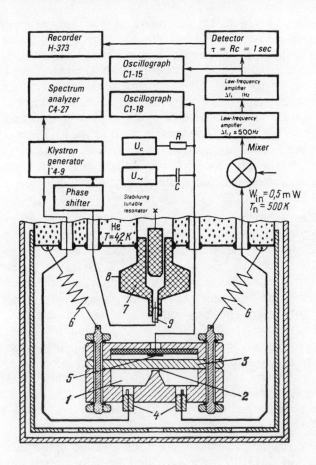

Fig. 29 Diagram of apparatus used to measure small mechanical vibrations. 1, electromagnetic resonator; 2, gap of the measuring capacitor; 3, locking diaphragm; 4, inductive coupling loops; 5, plate of the calibration capacitor; 6, antiseismic filter; 7, superconducting sapphire tuning resonator; 8, superconducting coating; 9, capacitive coupling probe.

units of volts).

The superconducting resonator and calibrating capacitor were suspended in a vacuum cryostat by a seismic-isolation filter (6 in figure 29) with a natural frequency of 3 Hz. This filter substantially attenuated the influence of external seismic and acoustic disturbances at the mechanical frequency f_m.

The pump oscillator that drove the readout resonator 1 was a reflex klystron stabilized by a tunable sapphire resonator 7 using frequency pulling (Braginsky, Panov, and Vasiliev 1981; Panov 1980). A coaxial line of equivalent electrical length $(2n+1)\lambda/4$ connected the klystron to the stabilizing resonator, terminating in a capacitive coupling probe 9 in the end of the resonator. The coupling was adjusted by the depth of the probe's insertion into the resonator. The coaxial line was coupled to the klystron at its other end by an inductance loop in the klystron resonator cavity. In addition, to provide phase control of the return signal from the stabilizing resonator, a phase shifter was connected in series with the coaxial line.

The stabilizing tunable resonator had a geometric factor of $\Gamma(\epsilon') = 80$ ohm and was fabricated from a monocrystal of sapphire covered with a superconducting lead coating; it had losses $\tan\delta \leqslant 1 \times 10^{-9}$, a frequency $f_r \approx 3$ GHz, and temperature $T = 2\,\mathrm{K}$. To tune the resonator's frequency, an opening 10 mm in diameter in the resonator's cavity was plugged by a sapphire rod 9 mm in diameter, and this rod was moved in and out. By this movement it was possible to alter the resonator's eigenfrequency over a range $\Delta f_r \approx 300$ MHz. This stabilizing resonator was placed in the cryostat along with the superconducting readout resonator and the calibrating capacitor, and their temperatures ($T = 4.2\,\mathrm{K}$) were maintained with the aid of a thermal conductor. At this temperature the quality factor of the stabilizing resonator was $Q_e^{st} = 2 \times 10^7$.

The regime of frequency pulling and the optimization of the stabilization factor S were provided by adjustments of the couplings and of the phase of the return signal from the stabilizing resonator. In this way, for the given quality factor of the superconducting resonator, it was possible to achieve the following parameters for the pump oscillator: the stabilization factor was $S \sim 10^4$ for an output power $W = 60$ mW, and the fractional frequency stability (Allan variance) was as small as $\sigma(\Delta\omega/\omega) \leq 3 \times 10^{-11}$ during 1 sec in the tuning frequency range. For these parameters, and at frequencies $\sim 8\,\mathrm{kHz}$ away from the carrier frequency, the spectral densities of the frequency and amplitude fluctuations were respectively

$$w_f \lesssim 3 \times 10^{-8} \text{ Hz}^2/\text{Hz} , \quad w_a \lesssim 2 \times 10^{-18} \text{ Hz}^{-1} .$$

For the given readout resonator parameters, these fluctuation levels allowed the diaphragm's vibration amplitude to be monitored down to $x = 2 \times 10^{-19}$ m in a bandwidth $\Delta f = 1$ Hz.

The readout resonator's microwave-frequency output signal at $f_p \pm f_m$ was monitored with a mixing device that had an effective noise temperature $T_n = 500$ K. A matching transformer and a low-noise, low-frequency, three-stage amplifier (with a total gain $K = 3 \times 10^8$ and noise temperature $T_n \approx 5$ K) were arranged at the output of the mixing device. The gain of the first stage was $K_1 = 300$ in a bandwidth $\Delta f_1 = 3$ kHz, that of the second was $K_2 = 10^3$ in $\Delta f_2 = 500$ Hz, and that of the third was $K_3 = 10^3$ in $\Delta f_3 = 1$ Hz. The bandwidth was narrowed to 1 Hz with a quartz filter adjusted to the signal frequency $f_m = 8$ kHz. This filtering near the frequency $f_m = 8$ kHz excluded from the record effects of the Brownian motion of the diaphragm, since the main part of the Brownian-motion spectrum was concentrated near the diaphragm's mechanical eigenfrequency $\simeq 40$ kHz. A linear detector with a relaxation time $\tau = 1$ sec followed the amplifier. A type H-373 recorder was used for recording the output signal and noise.

The adjustment of the system and the recording of small mechanical vibrations were performed in the following way. After the superconducting readout resonator was cooled down to its working temperature of $T = 4.2$ K, the resonator frequency and the quality factor were measured. Then the power from the pump oscillator was applied to the readout resonator through a mono-directional element, and the pump oscillator's frequency was adjusted to match the steepest point on the readout resonator's resonance curve. The power going through the resonator was measured at its output.

The pump power at the input of the mixing device was set at the level $W_{in} = 0.5$ mW, which produced the maximum signal/noise ratio for the entire monitoring system. Additional fine adjustments of the apparatus and the actual measurement of the mechanical vibrations were made with the calibration force set at a level that only slightly exceeded the intrinsic noise level of the monitoring system. The signal power at the input of the mixing device W_s was related to the amplitude x of vibration of the readout capacitor's diaphragm by

$$W_s = W_{in} \frac{Q_e^2}{4} \left(\frac{x}{d}\right)^2 .$$

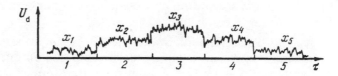

Fig. 30 Record of diaphragm's mechanical vibrations for several levels of calibration amplitude.

The minimum detectable amplitude of the diaphragm's vibration x_{min} was computed from the calibration force, which gave a barely detectable signal, and from the diaphragm's spring constant. To protect against errors in the value of the minimum detectable diaphragm displacement, all the parameters of the calibration system were measured experimentally and compared with their calculated values. The differences were small (no more than 15%) and thus supported the conclusion that the amplitude x_{min} determined from the calibration force was close to the experimental value.

An experimental record of the output amplitude of the monitoring system at frequency $f_m = 8$ kHz is shown in figure 30. There are five distinct parts of this record. Parts 1 and 5 show the intrinsic noise of the monitoring system when no calibration force was present ($x_1 = x_5 = 0$). Parts 2 and 4 show the output when the calibration amplitude was $x_2 = x_4 = 1.9 \times 10^{-18}$m. Part 3 is the record of vibrations with an amplitude of $x_3 = 3.8 \times 10^{-18}$m. A statistical analysis of the measurements shows that the minimum amplitude detectable x_{min} (one standard deviation σ) is equal to 6×10^{-19}m in a bandwidth $\Delta f = 1$ Hz and averaging time $\hat{\tau} = 1$ sec, and is 2×10^{-19}m for $\hat{\tau} = 10$ sec.

The sensitivity of this variable capacitance transducer was limited mainly by acoustic and seismic disturbances in the microwave-frequency circuits and by the finite noise temperature of the mixing device.

The sensitivity can be improved by increasing the quality factor of the transducer's resonator, by using low-noise circuits, and by improving the spectral characteristics of the pump oscillator. However, the experimental results for superconducting capacitive transducers with a small capacitance gap (Braginsky, Panov, and Popel'nyuk 1981) do not give great hope for a marked increase in sensitivity. Under actual conditions, sensitivity will be limited by a combination of the transducer's finite quality factor and the requirement that the electric field strength in its capacitance gap not exceed the breakdown field E_{br}. These produce the following fundamental limit on the minimum detectable amplitude x_{min}

$$\left(\frac{x}{d}\right)_{min} - \left(\frac{8kT_n}{\epsilon_0 E_{br}^2 V \omega_p Q_e \tau}\right)^{1/2}. \tag{10.1}$$

Here V is the effective volume of the capacitance gap of the transducer's resonator, Q_e is the resonator's quality factor, ϵ_0 is the dielectric constant, ω_p is the pump frequency, and τ is the duration of measurement. For the transducer described above, $E_{br} - 18\,MV\,m^{-1}$ (corresponding to $d - 10^{-5}m$), $\omega_p - 1.9 \times 10^{10}\,sec^{-1}$, $Q_e - 2 \times 10^6$, and $T_n - 500\,K$. Substituting these into equation (10.1) we obtain for the minimum measurable amplitude

$$x_{min} - 4 \times 10^{-20}\tau^{-1/2}m \,,$$

a factor of only 16 below the sensitivity achieved.

In conclusion, we note that the high sensitivity of this apparatus to small mechanical vibrations depends crucially on the use of two superconducting resonators: the capacitance transducer and the generator's stabilizing resonator.

Measurements of small electromagnetic losses in solids. The determination of the loss parameter $\tan\delta$ for a specimen of a chosen material is generally achieved by measuring its effect on the quality factor of a high-quality superconducting resonator. By this method, dielectric losses in liquid helium have been measured: $\tan\delta$ was found to be no greater than 10^{-10} at temperature $T - 4.2\,K$ and frequency ≈ 90 MHz (Mittag *et al.* 1973). Losses in dielectric solids have also been measured in this way, using either a cavity or helical resonator with an intrinsic quality factor of $Q_e \sim 10^7$ to 10^9. For instance, Hartwig and Grissom (1964) found $\tan\delta \approx 1.5 \times 10^{-4}$ for silicon at a frequency 98 MHz, and $\tan\delta \approx 1.2 \times 10^{-6}$ for Teflon at 118 MHz.

It should be noted that in a later measurement, Mittag *et al.* (1970) reported $\tan\delta$ for Teflon as no more than 4×10^{-7} at temperature $T - 4.2\,K$ and over a frequency range from 30 to 180 MHz. This is almost one order of magnitude smaller than the Hartwig-Grissom result. This discrepancy can be explained either

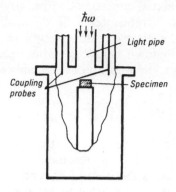

Fig. 31 Superconducting resonator for study of dielectric properties of semiconductors (Grissom and Hartwig 1966; Hinds and Hartwig 1971; Alworth and Haden 1971).

by a difference in the preparation of the specimens (surface contamination can cause additional losses) or by the presence of additional impurities in the interior of the specimen (impurities may cause large losses). Using superconducting resonators, Meyer (1977) investigated losses in polymers at low temperatures, finding values of $\tan\delta$ for polyethylene at $T = 2.2\,\mathrm{K}$ and frequency 6.9 GHz as small as $(3.7 \pm 0.05) \times 10^{-7}$. The high sensitivity of these methods of using superconducting resonators makes it possible to detect a density of defects of the order of 1 defect/10^{10} atoms and to investigate subtle effects in semiconductors caused by sunlight (Grissom and Hartwig 1966; Hinds and Hartwig 1971). These methods have produced many experimental data on the influence of light on the properties of semiconductors at low temperatures.

Figure 31 illustrates the standard scheme for this type of measurement. The specimen is placed in the capacitance gap of a helical or coaxial superconducting resonator. Capacitance probes or coupling loops are used to drive the resonator and extract its output signal. When the effects of light are studied, the light is supplied by a light pipe. For the supply of infrared radiation, a cooled tube is used instead of a light pipe. The measurement is taken either in a vacuum or in liquid helium at a temperature of 2 to 4.2 K. Analogous schemes have been used to measure the relaxation times and the lifetimes of free current carriers; the location of the Fermi level; the density, depth and occupation of traps;

and the capture cross-section for many semiconductors such as Si, CdS, GaAs, and CdTe.

An investigation of semiconductors at low temperatures by Alworth and Haden (1971) has shown that the dielectric constants of semiconductors change when they are bombarded by nuclear radiation. This effect is caused by the additional charge carriers that are created by the ionization of atoms, and it permits the use of superconducting resonators as sensitive radiation monitors and dosimeters. The frequency and quality factor of a superconducting resonator in which a radiation-exposed semiconductor is placed will be changed by amounts that depend on the exposure dosage.

Measurements of this type have been performed with super-conducting resonators at a frequency of $\sim$380 MHz. The CdS crystal under examination was exposed to a radioactive C^{14} source with an intensity of 10 and 1 μC. The dependence of the frequency shift on the exposure time was linear for small doses and corresponded to a shift rate of $\sim$4 kHz/min. With a radioactive source of 100 μC intensity, the initial shift rate was 100 Hz/min and it decreased constantly, changing by almost one order of magnitude over 1 hr. Thus, such a device can be used as a dosimeter for small radiation fluxes.

Studies of the properties of semiconductors have been carried out not only using superconducting cavity resonators, but also by measurements of tanδ in superconducting, dielectric resonators (see section 7) (Bagdasarov, Braginsky, and Zubietov 1977; Braginsky and Panov 1979; Braginsky, Panov, and Vasiliev 1981). Dielectric resonators are especially helpful in measuring very small dielectric losses because they are free of two key disadvantages of cavity reso-nators: changes of resonator parameters with time and interactions of the superconducting walls with residual gas.

To measure losses with a dielectric resonator, one can use the dielectric material to be studied as the interior of the resonator. One need only coat the surface of the dielectric sample with a superconducting layer to convert it into a resonator. If the resonator's quality factor is not restricted by the residual surface resistance, then from a measurement of the quality factor, one can infer the value of tanδ for the dielectric material. This method has allowed us to determine an upper limit for sapphire of tan$\delta \leq 10^{-9}$, and to find that tan$\delta \approx 8.2 \times 10^{-8}$ for Teflon at a frequency of 3.1 GHz and temperature $T = 2K$.

Measurements of the surface impedances of superconductors. Superconducting resonators have been used successfully to study the surface impedances of superconductors (Dheer 1961; Williams 1962). The surface resistance R_s (i.e., the active part of the surface impedance) is determined in most cases by measuring the quality factor of the resonator. If the quality factor, the oscillation mode, the resonator dimensions, and the coupling coefficients are known, the surface resistance R_s is easily determined.

Formally, the absolute value of the surface reactance X_s (i.e., the imaginary part of the surface impedance) is defined by the expression $X_s = 2\Gamma\Delta\omega/\omega_0$, where $\Delta\omega$ is the difference between the resonator frequencies of ideal (without skin effects) and real resonators. Practically speaking, however, this expression cannot be used because calculations of the ideal resonator frequency are inaccurate; errors in the resonator dimensions produce errors in ω_0 much greater than $\Delta\omega$. Therefore, only changes in the surface reactance ΔX_s due to changes of experimental parameters (e.g., temperature) can be determined by the resonator method. Such measurements also reveal the corresponding changes in the skin depth $\Delta\delta = \Delta X_s/\omega\mu_0$.

The possible applications of these methods are expanded by the use of resonators with sapphire interiors. In particular, detailed studies of the surface impedances of superconducting coatings are possible (Braginsky and Panov 1979; Balalykin, Zubietov, and Panov 1978; Braginsky, Panov, and Vasiliev 1981). To illustrate the experimental possibilities of this method, we shall describe measurements of this type in more detail.

It is known that both the real and imaginary parts of the surface impedance Z_s are finite at $T \neq 0$. In the important case where $T < 0.5T_c$, $\omega \ll kT/\hbar$, and $B \ll \mu_0 H_{c2}(T)$, the following expressions are valid

$$R_s = R_0 + R_s(T) + R_s(B, H^\omega),$$
$$X_s = X_0 + X_s(T) + X_s(B, H^\omega). \tag{10.2}$$

Here T and T_c are respectively the actual and critical temperatures of the superconductor, ω is the frequency of the resonator's electromagnetic field, B is the density of DC magnetic flux trapped in the superconductor, H^ω is the strength of the microwave-frequency magnetic field, and $H_{c2}(T)$ is the strength of the second critical field. The functions $R_s(T)$ and $X_s(T)$ characterize the temperature dependence of the surface impedance. R_0 and X_0 are independent and represent the residual surface resistance and reactance as determined by the limiting value of the skin depth, $\delta(T)$ at $T \to 0$.

There is a set of theoretical models (Abrikosov, Gorkov, and Khalatnikov 1958; Mattis and Bardeen 1958; Miller 1960; Maki 1965; Caroli and Maki 1967), based both on phenomenological ideas and on microscopic theory, that describes the dependences of R_s and X_s on temperature and magnetic field strength. Comparison of these models with detailed experimental results permits one to answer a variety of fundamental questions about the behavior of superconductors under various conditions.

In the measurements we will consider, superconducting coatings of lead and niobium were studied. Measurements of the surface impedance Z_s at a frequency of 2.89 GHz were performed, using sapphire resonators coated with lead or niobium (see section 7 for details). The resonators were placed either in liquid helium or in a high vacuum. A DC magnetic field $B \sim 10^{-5}$ to $2.5 \times 10^{-2}\,$T was created by coils outside of the cryostat. To reduce errors in the estimate of the resonator's demagnetization factor, measurements were performed for various directions of the external magnetic field relative to the axis of the resonator. The Earth's magnetic field was balanced to an accuracy of $\pm 5 \times 10^{-6}$T. The measured values of R_s for niobium coatings ($T_c = 9.22\,$K) and the predicted temperature dependence $R_s(T)$ according to the Mattis-Bardeen theory (Mattis and Bardeen 1958; Miller 1960) are shown in figure 32. For the parameter values $\Delta(0) = 1.85\,kT_c$, $v_F \approx 2.9 \times 10^5$m/sec, $l_e = 10^{-6}$m, $\delta(0) = 3.5 \times 10^{-8}$m, and $\xi = 3.9 \times 10^{-8}$m, the theoretical curve is close to the experimental results over a wide range of temperatures, even though the experimental value of the surface resistance at $T > T_c$, $R^{\mathrm{exp}} = 4.1 \times 10^{-3}$ohm is two times larger than the theoretical value. The high-temperature difference between R^{exp} and R is explained by the fact that at $T \geqslant T_c$ approximately 50% of the microwave power was lost to emission through the resonator walls because of the small thickness of the superconducting film ($d_f \approx 0.5\,\mu$m). This increased the experimental value of R^{exp} by a factor of two.

The measured values of $R_s(B)$ and $X_s(B)$ for lead and niobium coatings, with critical temperatures $T_c = 7.2\,$K and $T_c = 9.2\,$K, respectively, are shown in figure 33. The coatings used underwent a "fast" transition into the superconducting state in the

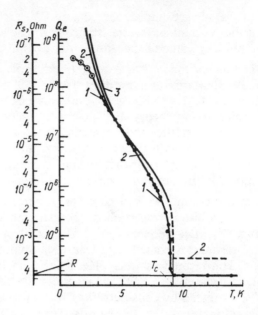

Fig. 32 Temperature dependence of Q_e and R_s for a niobium film (Balaly-kin, Zubietov, and Panov 1978). 1, experimental curve for niobium film with $T_c = 9.22\,\mathrm{K}$; 2, theoretical curve according to Mattis-Bardeen theory (Mattis and Bardeen 1958; Miller 1960); 3, experimental curve for massive specimen.

presence of the magnetic field. A fast transition was needed for full trapping of the external magnetic field by the superconductor; that is, the speed of the transition through the coating had to be greater than the speed of "creep" of the field (de Gennes 1966). The quantities $R_s(B)$ and $X_s(B)$ are functions of the magnetic flux density averaged over the resonator's surface, $\bar{B} = \alpha\mu_0\bar{H}$. The factor α depends on both the resonator geometry and the magnetic field orientation. For a resonator with approximately equal length and diameter and oscillation mode E_{010}, the factor α is nearly independent of the magnetic field direction, and its value is in the range $\alpha = 0.3 \pm 0.05$.

In addition to the measurements listed above, the critical field strength H_{c1}, at which the superconductivity is partially destroyed, was also measured. Allowing for the demagnetization factor of the resonator ($n = 0.8$), the values of the first critical magnetic field for

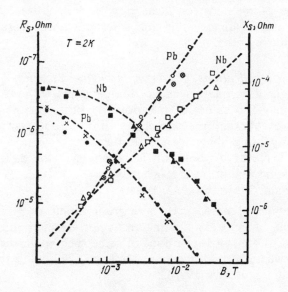

Fig. 33 Surface impedance as function of trapped magnetic field B for superconducting lead and niobium coatings (Balalykin, Zubietov, and Panov 1978).

the niobium coatings were $H_{c1} = 1.11 \times 10^5 \text{At/m}$ at $T = 4.2\,\text{K}$ and $H_{c1} = 1.27 \times 10^5 \text{At/m}$ at $T = 2\,\text{K}$. It should also be noted that when an initially strong magnetic field was removed, $R_s(B)$ was not reduced below $\approx 10^{-5}\text{ohm}$. This implies that there was residual magnetism associated with surface and structural defects in the superconducting coating.

The dashed lines in figure 33 are theoretical curves for $R_s(B)$ and $X_s(B)$. The agreement between the model (Balalykin, Zubietov, and Panov 1978) and the experimental data permits use of the model to calculate the effects of trapped magnetic fields on the quality factors and frequencies of coated resonators.

In particular, these data indicate that if one cools niobium-coated resonators in the presence of the Earth's magnetic field, the frequency reproducibility cannot be more than $\Delta\omega/\omega \approx 3 \times 10^{-9}$, and the quality factors will not exceed $Q_e \approx 10^9$.

We shall not describe here other details of the experimental studies of superconducting coatings, except to refer the reader to the results discussed in section 7. That section also contains a description of methods for measuring the quantities $\Delta X_s(T)$ and $\Delta X_s(H^\omega)$, which are essential for determining the parameters of superconducting resonators.

Other applications of superconducting resonators. High-Q electromagnetic resonators at low temperatures without superconductivity may also be useful in various physics experiments. As an example we note that in a test of the Ginzburg-Pitaevsky theory of superfluidity using a high-Q electromagnetic resonator, the measurements achieved a resolution for changes in the dielectric constant of liquid helium at the level $\Delta\epsilon/\epsilon \simeq 1 \times 10^{-9}$. (For details, see Panov and Sopyanin 1982, 1984.)

It should be noted that there are a variety of other possible applications for superconducting resonators beyond those already mentioned. Detailed descriptions of some such applications can be found elsewhere (see, for instance, Didenko 1973; and Mende, Bondarenko, and Trubitsyn 1976).

V   Mechanical Oscillators
in Physical Experiments

11. Mechanical Gravitational Antennae

The detection of gravitational waves is one of the most interesting problems in modern physics. Joseph Weber (1961) was the first to discuss the possibility of constructing earth-based gravitational antennae capable of monitoring gravitational-wave bursts from cosmic (including extragalactic) objects. Although Weber's first experiments (from 1969 to 1974) and other variants of them with similar sensitivities in many other laboratories failed to give positive results, second-generation gravitational antennae are presently being developed in twenty laboratories in many countries (see, for instance, Braginsky and Thorne 1983). These antennae are designed to detect gravitational-wave bursts originating in astrophysical catastrophes such as supernova explosions and black-hole and neutron-star collisions.

Fairly detailed theoretical analyses of various hypothetical sources of sporadic gravitational radiation have yielded remarkable results. Of particular interest is the conclusion that tens of percent of the mass involved in some catastrophic processes might be converted into gravitational radiation. Unfortunately, due to extreme theoretical and computational difficulties, and because electromagnetic (optical, radio, etc.) studies of the universe have given us little knowledge of the universe's strong-gravity features, the theoretical calculations give highly uncertain predictions for the characteristics of the gravitational waves bathing the earth. The astrophysical aspects of this problem and the possibility of acquiring a new channel of cosmic information have aroused considerable interest in the development of gravitational antennae; see, e.g., Dereulle and Piran (1983) and references therein. In this section we shall briefly describe the role of high-Q mechanical oscillators in the development of gravitational antennae.

Eight of the twenty laboratories just mentioned are developing mechanical antennae of the type proposed by Weber. Such an antenna is constructed from a massive cylinder and a high-sensitivity transducer that monitors low-mode quadrupole vibrations of the cylinder. Gravitational-wave bursts are expected to drive the cylinder's quadrupole vibrations. According to theory, some possible cosmic sources of gravitational radiation would provide a gravitational impulse of duration $\tau_{gr} \approx 10^{-3}$ to 10^{-4} sec with a waveform resembling a single period of a harmonic oscillation. To achieve optimal coupling of such a gravitational impulse to the antenna, the length of the cylinder L is chosen to be $L \approx v\tau_{gr}/2 \approx (0.5 \text{ to } 5)$ m, where v is the sound velocity in the cylinder. The change of the cylinder's vibration amplitude caused by the gravitational impulse is roughly

$$\Delta x_{gr} \approx h_{gr}L \approx \tfrac{1}{2}h_{gr}v\tau_{gr} . \qquad (11.1)$$

Here h_{gr} is the dimensionless amplitude of the general relativistic metric perturbation associated with the gravitational radiation.

Two estimates of the wave amplitude h_{gr} are commonly used: an optimistic one, $h_{gr} \approx 2 \times 10^{-19}$, and a more realistic one, $h_{gr} \approx 1 \times 10^{-21}$. The optimistic value corresponds to a supernova explosion at a distance of 3 megaparsecs, 100 times further than the outer limits of our galaxy, with 10% of its mass converted into gravitational waves (or a supernova in our own galaxy, with 0.001% mass conversion). The realistic value corresponds to a theoretical supernova explosion at a distance of 15 megaparsecs with 1% mass conversion (or one in our galaxy with 10^{-6}% mass conversion). Impulses with $h_{gr} \approx 2 \times 10^{-19}$ might conceivably reach the earth as often as once a month but probably are less frequent. Weaker impulses can be expected more frequently (for further details, see, for example, Thorne 1980; Smarr 1979).

It is clear from these estimates for h_{gr} that two or more antennae working in coincidence are necessary for gravitational wave experiments, and that extremely small amplitudes of vibration must be measured. The first, essential condition for the detectability of the antenna's wave-induced amplitude changes Δx_{gr} is that they must be no smaller than the thermal-noise fluctuations:

$$\Delta x_{gr} \approx \tfrac{1}{2} h_{gr} v \tau_{gr} \geqslant \sqrt{kT\hat{\tau}/M\omega Q} \; . \qquad (11.2)$$

Here M is the effective mass of the antenna (approximately one third of the actual mass of the cylinder for the lowest quadrupole vibration mode), Q is the quality factor for the chosen mode, $\hat{\tau}$ is the time taken to measure Δx, and ω is the eigenfrequency of the chosen mode, which should satisfy $\omega \tau_{gr} \approx 2\pi$. It is obvious that the measuring time $\hat{\tau}$ can be greater than τ_{gr}: the gravitational antenna, being a high-Q oscillator, "remembers" an impulsive excitation as long as $\tau^* = 2Q/\omega$.

Condition (11.2) can be rewritten in the form

$$Q \geqslant \frac{2kT}{\pi v^2 M h_{gr}^2} \frac{\hat{\tau}}{\tau_{gr}} \; . \qquad (11.3)$$

This inequality shows that both a decrease in T and an increase in M can compensate for a lower quality factor Q. This relationship has motivated two distinct directions of antenna development. One can use materials with a large expected quality factor Q and a comparatively small mass M (e.g., leucosapphire with M of order tens of kilograms), or one can use materials that have comparatively small quality factors (aluminum alloys or niobium), but that can be fabricated in massive (several ton) cylinders.

If we substitute into equation (11.3) the values $h_{gr} = 10^{-21}$, $T = 2K$, and $v = 1 \times 10^4$m/sec, we find that $Q \geqslant 2 \times 10^{10}\hat{\tau}/\tau_{gr}$ for $M = 10$kg, and $Q \geqslant 2 \times 10^8 \hat{\tau}/\tau_{gr}$ for $M = 10^3$kg. If $\hat{\tau} = \tau_{gr}$, the quality factors that have already been obtained for sapphire, or aluminum and niobium, are close to these requirements (see section 4 for details). Unfortunately, however, it is difficult to achieve $\hat{\tau} = \tau_{gr}$ because of noise in the sensor. Sensor noise is less serious for $\hat{\tau} \gg \tau_{gr}$; but the requirements for the quality factor are then much more strict. Further temperature reduction is one way to ease the requirement for the quality factor, and this route has been chosen by some experimenters.

A sensitive sensor for small vibrations [amplitude $\leq$ (1 to 2) $\times 10^{-19}$m] must be coupled to the antenna in such a way that the mechanical quality factor does not drop. But it is known that any sensing element affects the dynamical system that it measures and has a strong propensity to reduce the quality factor (see, for example, Braginsky and Manukin 1974). (This problem will not be discussed here, but we suggest the reader consult several publications describing ways of coupling the sensor to the mechanical resonator that have been adopted in various laboratories: Braginsky, Lanin, and Panov 1979; Blair *et al.* 1980a; Amaldi *et al.* 1980; and Paik 1980.)

Along with its dynamic, Q-reducing influence on the antenna, the sensor also exerts a fluctuating "back-action" force that produces amplitude fluctuations in the antenna and thereby reduces its sensitivity. To a certain extent, this fluctuating back action is due to practical effects and can be reduced. However, when the accuracy of the amplitude measurements is near the half-width of the wave packet for a coherent quantum state

$$\Delta x_{coh} = \sqrt{\hbar/2M\omega}, \tag{11.4}$$

the back action becomes quantum mechanical in origin and cannot be removed, even in principle.

This restriction (already discussed in section 2) shows that the disturbing back action of the sensor is not simply a practical difficulty to be overcome on the way to higher quality factors. On the contrary, there is a threshhold value of the quality factor above which the back action becomes unavoidable. In the case of continuous position measurements, this threshhold value is $Q \approx 2kT\hat{\tau}/\hbar$ (see Section 2), so that such measurements cannot reveal an amplitude change less than Δx_{coh}. Corresponding to this amplitude-measurement limit there is a "standard quantum limit" for the sensitivity of a mechanical gravitational antenna

$$h_{gr} \gtrsim \left[\frac{2\hbar}{M\omega(v\tau_{gr})^2}\right]^{\frac{1}{2}}. \tag{11.5}$$

For $M = 10^3$kg, $\omega\tau_{gr} \approx 2\pi$, and $v = 10^4$m/sec, the limiting sensitivity [equation (11.5)] is $h_{gr} \geq 3 \times 10^{-21}$. In other words, using an ordinary (in the classical sense) measuring system, it is not possible in principle, even with a very massive antenna, to measure gravitational impulses with the realistic 1×10^{-21} value of h_{gr}.

Actually, the standard quantum limit [equation (11.5)] is an unavoidable consequence of the Heisenberg uncertainty principle only when the sensing system makes continuous measurements of the mechanical oscillator's position. The issue of how one might circumvent this limit by other types of couplings of the sensor to the antenna ("quantum nondemolition measurements") has been addressed by Braginsky, Vorontsov, and Khalili (1978), Thorne *et al.* (1978), and Hollenhorst (1979). One solution to the problem, which is simple in principle but difficult in practice, is to use a sensor that measures only one of the two quadrature components, X_1 or X_2, of the antenna's position

$$x(t) = X_1\sin\omega t + X_2\cos\omega t . \qquad (11.6)$$

If, for instance, the X_1-component is measured, then the X_2-component will be disturbed by an amount inversely proportional to the accuracy of the X_1-component measurement, since the product $\Delta X_1 \Delta X_2$ must satisfy the Heisenberg uncertainty principle

$$\Delta X_1 \Delta X_2 \geqslant \frac{\hbar}{2M\omega} .$$

This simple idea has not been realized in practice as yet, although apparently viable sensor designs have been developed and analyzed. One promising design needs for its practical realization a parametric electromechanical transducer (analogous to that described in Section 10) with a very high electromagnetic quality factor. If the quality factor Q_e is greater than the ratio ω_e/ω, then such a transducer with an appropriate type of high-frequency pumping will achieve a measurement of X_1 with accuracy $\Delta X_1 < \Delta x_{\text{coh}}$ and a corresponding gravitational-wave sensitivity $h_{gr} <$ (expression 11.5) (Panov and Khalili 1980; Caves, Drever, and Thorne 1980; Braginsky, Vorontsov, and Thorne 1980).

In summary, it should be noted that research on systems with small dissipation that is being carried out as part of the gravitational-antennae-development program is producing ideas, techniques, and apparatus that will be useful outside the field of gravitation physics.

12. Applications of High-quality Mechanical Resonators to Frequency Stabilization

Here we will consider the use of high-quality mechanical resonators as the key elements in frequency-stabilization systems. Frequency stabilization by vibrating mechanical elements is widely used in science and engineering because the eigenfrequencies of mechanical elements are more independent of external factors than are the parameters of electric circuits. Stabilizing resonators must have a small frequency drift with temperature, a high quality factor (which is of great importance in holding the frequency stable in the

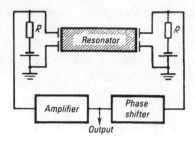

Fig. 34 Schematic diagram of electromechanical generator.

presence of random variations of electric circuit parameters), and minimum aging, that is, minimum systematic, uncontrolled changes of frequency with time.

Quartz-crystal resonators satisfy these requirements. The piezoelectric effect in quartz makes it easy to convert the crystal's vibrations into electric signals. The long-term drift $\Delta\omega/\omega$ of a self-excited oscillator stabilized by a precision quartz resonator is not more than 10^{-10}/month (Smagin 1977). In one experiment (Smagin 1975), an oscillator stabilized with a quartz resonator was cooled down to $T = 4.2\,\text{K}$. The oscillator had a quality factor of $Q \approx 10^9$ and a temperature frequency coefficient of $3 \times 10^{-10}\text{K}^{-1}$. The short-term frequency stability of the oscillator was 4×10^{-14} for $10^2\,\text{sec}$, which is comparable to that of quantum frequency standards.

We will not discuss the use of quartz resonators any further since there is already a voluminous literature on the subject (see, for example, Smagin and Yaroslavsky 1970). Instead we will use sapphire mechanical resonators to illustrate the main features of self-excited oscillators stabilized by high-Q mechanical resonators.

Mechanical sapphire resonators with input and output transducers were considered in section 5. Such a configuration can be regarded as an electromechanical resonance element with electrical input and output. If the signal at the output of such a resonator is amplified and fed back with appropriate phase at the input (figure 34), the result is a closed circuit that can undergo self-excited electromechanical oscillations; that is, an "electromechanical generator."

A mechanical resonator (for example, a cylindrical one), can be thought of as a distributed system with an infinite set of vibrational modes. It is possible to satisfy the self-excitation conditions for the first vibrational mode of the resonator alone and avoid excitations of the other modes. Then the displacement x of the ends of the resonator obeys the equation (cf. Section 5)

$$\ddot{x} + 2\theta\,\dot{x} + \omega_0^2 x = \frac{\beta_1\beta_2\beta_3}{M}\,(x_\tau - \gamma_e x_\tau^3)\,, \qquad (12.1)$$

where M, ω_0, and θ are the equivalent mass, eigenfrequency, and damping constant of the resonator, respectively; β_1 is the conversion factor of the capacitance sensor at the resonator output, β_2 is the gain factor of the amplifier, and β_3 is the factor for the conversion of electric signals at the resonator input into the force that excites the resonator vibrations. The subscript τ signifies that the electrical signal in the circuit that excites the resonator is retarded in phase by $\varphi = \omega\tau$ relative to the displacement of the ends of the resonator; that is, $x_\tau(t) \equiv x(t - \tau)$. The quantity γ_e describes the nonlinearity of the electrical circuit.

Bogolyubov and Mitropolsky (1963) give a method for solving equation (12.1) with two successive approximations. In the first approximation, the expressions for the amplitude x_0 and the frequency ω are the following:

$$x_0^2 = \frac{2\theta\,\omega + \beta\sin\varphi}{(3/4)\beta\gamma_e\sin\varphi}\,, \qquad (12.2)$$

$$\omega = \omega_0 + \frac{\theta}{\tan\varphi}\,, \qquad (12.3)$$

where $\beta \equiv \beta_1\beta_2\beta_3/M$.

From equations (12.2) and (12.3) we can find the minimum amplifier gain factor β_2 that is required for self-excited oscillations (that is, for $x_0^2 > 0$). When expressed in terms of the resonator parameters this minimum gain is

$$\beta_2 = \frac{4M\omega^2 d^3}{\epsilon_0 S U_c^2 Q}\,.$$

For $M = 0.5\,\text{kg}$, $\omega = 2 \times 10^5\,\text{sec}^{-1}$, $d \equiv$(the distance between capacitor plates and the end of the resonator)$= 10^{-3}\,\text{m}$, $S \equiv$(the area of the plates)$= 4 \times 10^{-4}\,\text{m}^2$, $U_c \equiv$(constant voltage applied to the plates)$= 100\,\text{V}$, and $Q \equiv$(mechanical quality factor)$= 5 \times 10^9$, we find that $\beta_2 = 500$ is required for self-excited oscillation.

Equation (12.3) shows that the optimal phase shift in the oscillator circuit is $\pi/2 + 2\pi n$. In this case, the generated frequency is determined solely by the eigenfrequency of the oscillator. The second approximation gives a fractional correction to the frequency of order Q^{-2}. This correction arises because the nonlinearity of the electrical circuit gives rise to harmonics and a corresponding slight shift of the fundamental frequency.

Using equations (12.2) and (12.3), one can analyze the principal sources of fluctuations in the generator's frequency. For quasistatic parameter variations, the frequency change is given, to first order in Q^{-1}, by

$$\frac{\Delta\omega}{\omega} = \frac{\Delta\omega_0}{\omega_0} + \tfrac{1}{2}Q^{-1} \frac{1}{\sin^2\varphi} \Delta\varphi + Q^{-1}\cot\varphi \frac{\Delta Q}{Q} . \qquad (12.4)$$

It is clear from this equation that a high resonator quality factor plays a crucial role in reducing the influence of drifts in the oscillator circuit on the oscillator frequency. Numerical estimates of the resulting oscillator frequency drift can be found in Braginsky and Mitrofanov (1978).

A typical feature of such an electromechanical generator is the very weak coupling between the mechanical resonator and the electrical circuit that produces the feedback signal. That is why a large amplifier gain β_2 is needed. As a result, a major source of the fluctuations that cause frequency drift is noise in the electromechanical transducer, specifically the thermal noise of its input resistance, $R_{\text{in}} = \text{Re}(Z_{\text{in}})$.

The effect of this thermal noise can be computed by augmenting the equation of motion (12.1) with the appropriate fluctuating force term. The resulting equation is well known and can be solved by standard methods (e.g., Malakhov 1968). The resulting natural linewidth is

$$(\Delta\omega)_{\text{nat}} = \frac{2\pi^2 kT\omega^2 \text{Re}Z_{\text{in}}}{x_0^2\beta_1^2 Q^2} , \qquad (12.5)$$

where x_0 is the resonator vibration amplitude and β_1 is the capacitance sensor's conversion factor. Substituting in equation (12.5) $\text{Re}Z_{\text{in}} = 3 \times 10^5 \text{ohm}$ and $x_0 = 10^{-8}\text{m}$, we find $(\Delta\omega)_{\text{nat}} = 10^{-15}\text{sec}^{-1}$.

The vibration power dissipated in the resonator for a vibration amplitude x_0 is

$$W = \frac{M\omega^3 x_0^2}{2Q}.$$

For the parameters used above, $W \approx 4 \times 10^{-11}$W, and the signal power at the output of the capacitance sensor is $W_e = 3 \times 10^{-12}$W.

This small signal power in the sensor gives its thermal noise considerable influence on the oscillator frequency. This noise is added to the signal, causing signal frequency fluctuations

$$\left[\frac{\Delta\omega}{\omega}\right]_{ad} = \frac{1}{\omega\hat{\tau}} \sqrt{W_n/W_e},$$

where $W_n = kT\Delta\omega_f/2\pi$ is the noise power in the frequency band $\Delta\omega_f$, and $\hat{\tau}$ is the averaging time. For $\Delta\omega_f = 6 \times 10^2 \text{sec}^{-1}$, the signal frequency fluctuations due to the additive noise are $(\Delta\omega/\omega)_{ad} = 2 \times 10^{-9}\hat{\tau}^{-1}$. To reduce this value further, one must narrow the frequency band $\Delta\omega_f$ of the filter through which the resonator signal passes.

One of the most crucial factors affecting the generator's frequency is variation of the resonator temperature. The temperature coefficients of the elasticity constants of sapphire all have the same sign (Ginnigs and Furukawa 1953); therefore, there is no temperature region with a zero temperature frequency coefficient. But the temperature dependences of the elastic modulae and linear expansion coefficient are much weaker at low temperatures than at room temperature. For instance, at $T = 4.2$K the value of the temperature frequency coefficient for a resonator with longitudinal vibrations along the L_2-axis is determined by the temperature dependence of the elastic modulae and is approximately equal to 3×10^{-10}K^{-1} (Bagdasarov, Braginsky, and Mitrofanov 1974). Thus, if the resonator's temperature is held fixed to an accuracy of $\Delta T/T \approx 3 \times 10^{-4}$ at temperature $T = 4.2$K, then its eigenfrequency can be maintained with an accuracy of $\Delta\omega_r/\omega_r \approx 10^{-13}$.

It is known that the stability of a quartz resonator is limited primarily by slow systematic changes of its parameters, that is, by resonator "aging." Although a large amount of experimental data has been accumulated on the aging of quartz resonators, there is today no satisfactory model that accounts quantitatively for the aging (Smirnov 1973).

Lowering the resonator temperature slows the processes that cause irreversible changes in resonator properties and, consequently, slows the drift of the eigenfrequency. Resonator frequency drift is associated with changes in both its mass and its rigidity. The mass change is caused mainly by adsorption and desorption processes on the crystal surface. The desorbed mass consists of gas molecules, molecules of other compounds dissolved in the crystal, and bits of crystal that are weakly bonded to the crystal surface. According to Frenkel, the time that an adsorbed molecule stays on the crystal surface is

$$\tau_a = T_a \exp\frac{\varepsilon}{RT} , \qquad (12.6)$$

where ε is the bond energy (which depends on the properties of the adsorbed molecule and the surface), R is the universal gas constant, T is the surface temperature, and T_a is the period of oscillation of the adsorbed molecule.

Equation (12.6) shows that the desorption process is considerably weakened at liquid-helium temperatures. At such temperatures only helium molecules contribute to the adsorption-desorption equilibrium because the partial vapor pressure of other substances is very small. A monomolecular film of helium on the resonator surface will change its eigenfrequency by $\sim 10^{-9}$. At a temperature of 4.2 K, the change in the adsorbed film is very slow, and therefore, the resulting frequency drift is very small.

The increase in the rigidity of a resonator due to suspension by thin threads or wire has been estimated by Braginsky and Mitrofanov (1978). The resulting frequency drift, depending on the type of suspension used, is no larger than 10^{-12}.

Most of these influences on the frequency stability of an electromechanical generator have practical rather than fundamental origins, and the numerical estimates given here are for normal laboratory working conditions.

There exists an ultimate quantum mechanical limit for the frequency stability of a mechanical resonator, analogous to that found for electromagnetic oscillators as discussed in chapter 4. The influence of thermal fluctuations can be greatly reduced for a resonator with an optimally designed and coupled sensor by lowering the temperature, but the back action of the sensor cannot be eliminated completely. For a continuous sensor reading (required in order to determine the phase and thus the frequency of oscillation), the resonator's position cannot be measured with an accuracy better than the coherent-state halfwidth

$$\Delta x = \left[\frac{\hbar}{2M\omega} \right]^{1/2} .$$

This results in the following estimate for the limiting frequency stability

$$\left[\frac{\Delta\omega}{\omega} \right]_{\text{lim}} = \left[\frac{\hbar}{2M\omega^2 x_0^2 \hat{\tau}} \right]^{1/2} , \qquad (12.7)$$

where it is assumed that the measurement time $\hat{\tau}$ is much shorter than the relaxation time $\tau_* = 2Q/\omega$ of the resonator.

Equation (12.7) for the limiting stability can be rewritten in terms of the geometric and physical properties of the resonator by introducing the dimensionless vibration amplitude $\xi = x_0/L$ and expressing the resonator's eigenfrequency in terms of its mass M, volume V, length L, and Young's modulus Y

$$\left[\frac{\Delta\omega}{\omega} \right]_{\text{lim}} = \frac{2}{\pi\xi} \left[\frac{\hbar}{YV\hat{\tau}} \right]^{1/2} \approx 5 \times 10^{-15}\hat{\tau}^{-1/2} .$$

The numerical value is for the sapphire resonator considered above.

The frequency stability of an electromechanical generator with a sapphire mechanical resonator has been studied experimentally by Apalkov, Mitrofanov, and Shiyan (1978). A longitudinally vibrating sapphire resonator with an eigenfrequency of 38.1 kHz and a quality factor of 3.9×10^9 was placed in a helium cryostat at 4.2 K. The electrical circuit, that is, the amplifier and phase shifter, was outside the cryostat. Equation (12.4) shows that the requirement for the phase stability of the amplifier was quite modest. A phase drift $\Delta\varphi < 10^{-2}$ in the amplifier would cause a fractional frequency change of the generator no greater than 2×10^{-12}. Therefore, a conventional amplifier could be used in the feedback circuit. Its gain factor under normal working conditions was $\beta_2 \approx 10^3$. The steady-state amplitude of the resonator's vibrations was $x_0 \approx 10^{-8}$m, which corresponded to a power dissipation in the resonator $W = 10^{-10}$W. The resonator quality factor changed by no more than $\approx 10\%$ during the experiment.

The experiment involved comparing the frequency of the sapphire-stabilized generator with the standard signal of a radio station broadcasting at 66.(6) kHz and having a stability of $\Delta\omega/\omega < 5 \times 10^{-12}$ (State Standard of Time and Frequency, 1978). The generator's frequency was adjusted by changing the resonator

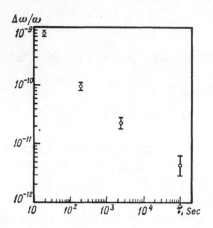

Fig. 35 Fractional frequency stability as function of averaging time for electromechanical frequency generator stabilized by mechanical sapphire resonator.

dimensions, so that when the generator frequency (multiplied by an integer, 7) was mixed with the broadcast frequency (multiplied by an integer, 4), the frequency difference was several cycles per second. This method provided the frequency resolution necessary for an accurate comparison of the frequencies. The difference between the frequencies was measured with a frequency meter synchronized to the broadcast signal. No special efforts were made in the experiment to stabilize the temperature of the liquid helium or the amplitude of the resonator's vibration.

Figure 35 gives the results of this experimental study of the frequency stability of an electromagnetic generator stabilized with a sapphire resonator. The rms value of the fractional frequency fluctuation $\Delta\omega/\omega$ is shown as a function of averaging time. For an averaging time of 1 day, $\Delta\omega/\omega$ was 4.4×10^{-12}. These fluctuations were evidently due primarily to phase distortion of the propagating radio signal and man-made interference. The noise in the generator's transducer and amplifier contributed significantly to the frequency fluctuations only for averaging times shorter than 1 day. No systematic frequency shift, except a diurnal one, was observed during the 4 days of measurement.

The level of stability obtained in this experiment is clearly not the limiting stability for electromechanical generators stabilized with high-Q mechanical sapphire resonators. In fact, even the limiting sensitivity of the specific generator used in this experiment is unknown, because the generator was compared with a frequency standard of inadequate stability.

Epilogue to the English Edition

The authors express their sincere gratitude to Professor R. F. C. Vessot, who urged them to write the following philosophical epilog for the English edition of this book.

An unbiased observer (a theorist) will likely agree that the community of experimental physicists is faced today with a rather long list of unsolved, fundamental physical problems that entail the measurement of very weak forces. A partial list includes:

1. The detection and study of bursts of gravitational radiation from astrophysical catastrophes in our Galaxy and in distant galaxies.

2. The detection and study of the relic background gravitational radiation produced in the early universe (analog of the cosmic electromagnetic background radiation).

3. The detection and study of the cosmic neutrinos, with nonzero rest mass, which are thought to pervade our galaxy with a density in the range 10^2 to $10^8 \mathrm{cm}^{-3}$.

4. The detection and measurement of the neutrino magnetic moment.

5. The detection and measurement of the gravitomagnetic field of the earth (i.e., the dragging of inertial frames by the earth's rotation), as predicted by general relativity.

6. The measurement of nonlinear gravitational effects predicted by general relativity.

The authors of this book are sure that various types of high-Q electromagnetic or mechanical resonators will be used, directly or indirectly, in the experiments which tackle these problems; and it is very likely that the feasibilities of some of these experiments will depend on the possibility of reaching a definite level of the quality factor. This is why we regard the struggle to isolate one degree of freedom from all others — i.e., the struggle for high Q — as a noble task for experimenters.

There is another field in experimental physics in which high quality factors are extremely important: Quantum nondemolition measurements. The feasibility of such measurements (or, at least, of some variants of such measurements) depends on a combination of strong nonlinearities and high Q. To our knowledge theorists until now have been unable to answer the question: Do the laws of physics demand any connection between the quality factor of an oscillator and its level of nonlinearity?

We conclude with a final, very philosophical remark: Any installation (experimental device) is wiser than the experimenter who created it. Due to this wisdom, the installation often offers many sophisticated questions to the experimenter — and sometimes gives him "wrong" answers. The experimenter may sometimes win the resulting battle with his installation, but he is likely to win only if he keeps in mind *all* the relevant knowledge about experimental physics accumulated by previous generations of experimenters. This is why it may be worthwhile for him or her to absorb the lore described in the present book.

Bibliography

Abragam, A., and Bleaney, B. 1970. *Electron Paramagnetic Resonance of Transition Ions*. Oxford: Clarendon.

Abrikosov, A. A. 1965. *Uspekhi Fiz. Nauk*, **87**, 125.

Abrikosov, A. A., and Khalatnikov, I. M. 1959. *Low Temperature Physics*. Moscow: IL.

Abrikosov, A. A., Gorkov, L. P., and Khalatnikov, I. M. 1958. *Zh. Eksp. Teor. Fiz.*, **35**, 265.

Agyenman, K., Puffer, I. M., Yasaitis, J. A., and Rose, R. M. 1977. *IEEE Trans. Magn.*, **MAG-13**, 343.

Akhiezer, A. Y. 1938. *Zh. Eksp. Teor. Fiz.*, **8**, 1318.

Akhmatov, A. S. 1963. *Molecular Physics of Boundary Friction*. Moscow: Fizmatgiz.

Allan, D. W. 1966. *Proc. IEEE*, **54**, 221.

Allen, M. A., Farkas, Z. D., Hogg, H. A., Hoyt, E. W., and Wilson, P. B. 1971. *IEEE Trans. Nucl. Sci.*, **NS-18**, 168.

Altman, J. 1964. *Microwave Circuits*. Princeton: Van Nostrand.

Altshuler, S. A., and Kozirev, B. M. 1972. *Electron Paramagnetic Resonance*. Moscow: Nauka.

Alworth, C. W., and Haden, C. R. 1971. *J. Appl. Phys.*, **42**, 166.

Amaldi, E., Modena, I., Pallotino, G. V., Pizzella, G., Ricci, F., Bonifazi, P., Fulgini, F., Giovanardi, U., Iafolla, V., and Ugazio, S. 1980. In *Abstracts of 9th Int. Conf. Gen. Rel. Grav.* (Jena, DDR), p. 410.

Anderson, O., and Bömmel, H. 1955. *J. Amer. Ceram. Soc.*, **38**, 125.

Apalkov, V. K., Mitrofanov, V. P., and Shiyan, V. S. 1978. *Doklady Akad. Nauk*, **242**, 578.

Arnolds, G., and Proch, D. 1977. *IEEE Trans. Magn.*, **MAG-13**, 500.

Arecchi, F.T. and Schulz-DuBois, E.O., eds. 1972. *Laser Handbook*. Amsterdam: North-Holland.

Bagaev, S. N., Baklanov, E. V., and Chebotaev, V. N. 1972. *Pisma v. Zh.*

Eksp. Teor. Fiz., **16**, 344.

Bagdasarov, Kh. S., Braginsky, V. B., and Mitrofanov, V. P. 1974. *Kristallografiya,* **19**, 883.

Bagdasarov, Kh. S., Braginsky, V. B., and Zubietov, P. I. 1977. *Pisma v. Zh. Tekh. Fiz.*, **3**, 991.

Bagdasarov, Kh. S., Braginsky, V. B., Mitrofanov, V. P., and Shiyan, V. S. 1977. *Vestn. Mosk. Univ., Ser. Fiz. Astron.*, No. 9, 98.

Balalykin, N. I., Zubietov, P. I., and Panov, V. I. 1978. *Pisma v. Zh. Tekh. Fiz.*, **4**, 407.

Belyaev, A. M., ed. 1974. *Ruby and Sapphire.* Moscow: Nauka.

Benaroya, R., Clifft, B. E., Johnson, K. W., Markovich, P., and Wesolowski, W.A. 1975. *IEEE Trans. Magn.,* **MAG-11**, 413.

Biquard, F., and Septier, A. 1966. *Nucl. Instrum. & Methods,* **44**, 18.

Biquard, F., Grivet, P., and Septier, A. 1968. *IEEE Trans. Instr. and Meas.,* **17**, 354.

Blair, D. G., and Jones, S. 1985. *IEEE Trans. Magn.,* **MAG-21**, 142; also *J. Appl. Phys.* (to be submitted).

Blair, D. G., Bryant, J. A., Buckingham, M. J., Davidson, J. A., Edwards, C., Ferreirinho, J., Griffiths, K. D., James, R. N., van Kann, F., Mann, L. D., and Veitch, P. J. 1980a. In *Abstracts of 9th Int. Conf. Gen. Rel. Grav.* (Jena, DDR), p. 411.

Blair, D. G., Buckingham, M. I., Edwards C., Ferreirinho, J., James, R. N., Mann, A. G., van Kann, F. J., and Veitch, P. 1980b. In *Proc. 2nd Marcel Grossman Conf. on General Relativity,* ed. R. Ruffini. Amsterdam: North-Holland.

Bogolyubov, N. N., and Mitropolsky, U. A. 1963. *Asymptotic Methods in the Theory of Non-Linear Oscillations.* Moscow: Fizmatgiz.

Bömmel, H. E., and Dransfeld, R. 1960. *Phys. Rev.,* **117**, 1245.

Bordoni, P. C. 1947. *Nuovo Cim.,* **4**, 177.

Braginsky, V. B., and Khalili, F. Ya. 1980. *Zh. Eksp. Teor. Fiz.,* **78**, 1712.

Braginsky, V. B., and Manukin, A. B. 1974. *Measurement of Weak Forces in Physics Experiments.* Moscow: Nauka; English translation, Chicago: University of Chicago Press, 1977.

Braginsky, V. B., and Mitrofanov, V. P. 1978. *Vestn. Mosk. Univ., Ser. Fiz. Astron.,* No. 4, 45.

Braginsky, V. B., and Nazarenko, V. S. 1969. *Zh. Eksp. Teor. Fiz.,* **57**, 1431.

Braginsky, V. B., and Panov, V. I. 1979. *IEEE Trans. Magn.,* **MAG-15**, 30.

Braginsky, V. B., and Thorne, K. S. 1983. In *Proc. 9th Int. Conf. Gen. Rel. Grav.,* ed. E. Schmutzer. Cambridge: Cambridge Univ. Press, p. 239.

Braginsky, V. B., and Vyatchanin, S. P. 1980. *Doklady Akad. Nauk,* **252**, 584.

Braginsky, V. B., and Vyatchanin, S. P. 1978. *Zh. Eksp. Teor. Fiz.,* **74**,

828.
Braginsky, V. B., Lanin, Yu. B., and Panov, V. I. 1979. *Vestn. Mosk. Univ., Ser. Fiz. Astron.,* No. 5, 87.

Braginsky, V. B., Panov, V. I., and Timashev, A. V. 1981. *Doklady Akad. Nauk,* **267,** 74.

Braginsky, V. B., Panov, V. I., and Vasiliev, S. I. 1981. *IEEE Trans. Magn.,* **MAG-17,** 955.

Braginsky, V. B., Vasiliev, S. I., and Panov, V. I. 1980. *Pisma v. Zh. Eksp. Teor. Fiz.,* **6,** 665.

Braginsky, V. B., Vorontsov, Yu. I., and Khalili, F. Ya. 1978. *Pisma v. Zh. Eksp. Teor. Fiz.,* **27,** 296.

Braginsky, V. B., Vorontsov, Yu. I., and Khalili, F. Ya. 1977. *Zh. Eksp. Teor. Fiz.,* **73,** 1340.

Braginsky, V. B., Vorontsov, Yu. I., and Thorne, K. S. 1980. *Science,* **209,** 547.

Braginsky, V. B., Vyatchanin, S. P., and Panov, V. I. 1979. *Doklady Akad. Nauk,* **247,** 583.

Braginsky, V. B., Bagdasarov, Kh. S., Panov, V. I., and Ilchenko, V. S. 1985. *Uspekhi Fiz. Nauk,* **145,** 151.

Brugger, K. 1964. *Phys. Rev. A.,* **144,** 1611.

Cady, W. G. 1946. *Piezoelectricity: An Introduction to the Theory and Applications of Electromechanical Phenomena in Crystals.* New York: McGraw-Hill.

Caroli, C., and Maki, K. 1967. *Phys. Rev.,* **159,** 316.

Caves, C. M., Drever, R., and Thorne, K. S. 1980. In *Abstracts of 9th Int. Conf. Gen. Rel. Grav.* (Jena, DDR), p. 424.

Dammertz, G., Hahn, H., and Halbritter, J. 1971. *IEEE Trans. Nucl. Sci.,* **NS-18,** 153.

de Gennes, P. G. 1966. *Superconductivity of Metals and Alloys.*

De Goes, R. M., and Dreyfus, B. 1967. *Phys. Stat. Sol.,* **22,** 77.

Dereulle, N., and Piran, T., eds. 1983. *Gravitational Radiation.* Amsterdam: North-Holland.

Dheer, P. N. 1961. *Proc. Roy. Soc. A,* **260,** 333.

Dick, G. J., and Strayer, D. M. 1984. In *Proc. 38th Annual Frequency Control Symposium,* Philadelphia (in press).

Didenko, A. I. 1973. *Superconducting Waveguides and Resonators.* Moscow: Sov. Radio.

Diepers, H., and Martens, H. 1972. *Phys. Lett. A,* **38,** 337.

Diepers, H., Schmidt, O., Martens, H., and Sun, F. S. 1971. *Phys. Lett. A,* **37,** 139.

Di Nardo, A. J., Smith, J. C., and Abrams, F. S. 1971. *J. Appl. Phys.,* **42,** 186.

Dobrosmyslov, V. S., and Vzyatyshev, V. F. 1973. *Trudy MEI,* **161,** 78.

Dodonov, V. V., Manko, V. I., and Rudenko, V. N. 1980. *Kvant. Elektron.,* **7,** 2124.

Fairbank, W. M., and Schwettman, H. A. 1967. *Cryog. Eng. News,* **2,**

No. 8, 46.

Fine, M. E., Van Duyne, H., and Kenney, N. T. 1954. *J. Appl. Phys.,* **25,** 402.

Flecher, P., Halbritter, J., Hietschold, R., Kneisel, P., Kühn, W., and Stoltz, O. 1969. *IEEE Trans. Nucl. Sci.,* **NS-16,** 1018.

Ginnigs, D. C., and Furukawa, T. 1953. *J. Amer. Chem. Soc.,* **75,** 522.

Ginzburg, V. L., and Landau, L. D. 1950. *Zh. Eksp. Teor. Fiz.,* **20,** 1069.

Giordano, S., Hahn, H., Halama, H. J., Luhman, T. S., and Bauer, W. 1975. *IEEE Trans. Magn.,* **MAG-11,** 437.

Golovashkin, A. I., Levchenko, I. S., and Motulevich, G. P. 1975. *Trudy FIAN,* **82,** 72.

Golovashkin, A. I., Levchenko, I. S., and Motulevich, G. P. 1972. *Fizika Metallov Melallovedenie,* **33,** 1213.

Gorelik, G. S. 1959. *Oscillations and Waves.* Moscow: Fizmatgiz.

Grissom, D., and Hartwig, W. H. 1966. *J. Appl. Phys.,* **37,** 4784.

Gurevich, V. L. 1980. *Kinetics of Phonon Systems.* Moscow: Nauka.

Gurevich, V. L. 1979. *Fizika Tverdogo Tela,* **21,** 345.

Gusev, A. V., and Rudenko, V. N. 1978. *Zh. Eksp. Teor. Fiz.,* **74,** 819.

Gusev, A. V., and Rudenko, V. N. 1976. *Radiotekhn. i. Electron.,* **21,** 1865.

Gvozdover, S. D. 1956. *Theory of Microwave Electronic Devices.* Moscow: Fizmatgiz.

Hahn, H., Halama, H. J., and Foster, E. H. 1968. *J. Appl. Phys.,* **39,** 2606.

Halama, H. J. 1971. *IEEE Trans. Nucl. Sci.,* **NS-18,** 188.

Halbritter, J. 1975. *IEEE Trans. Magn.,* **MAG-11,** 427.

Halbritter, J. 1972. *Part. Accel.,* **3,** 163.

Halbritter, J. 1971. *J. Appl. Phys.,* **42,** 82.

Harrison, W. A. 1970. *Solid State Theory.* New York: McGraw-Hill.

Harrop, P. J., and Creumer, R. H. 1963. *Brit. J. Appl. Phys.,* **14,** 335.

Hartwig, W. H. 1973. *Proc. IEEE,* **63,** 58.

Hartwig, W. H., and Grissom, D. 1964. In *Proc. Int. Conf. on Low Temp. Phys.* New York: Plenum Press, Pt. A, 1243.

Hillenbrand, B., and Martens, H. 1976. *J. Appl. Phys.,* **47,** 4151.

Hillenbrand, B., Martens, H., Pfister, H., Schnitzke, K., and Uzel, Y. 1977. *IEEE Trans. Magn.,* **MAG-13,** 491.

Hillenbrand, B., Martens, H., Pfister, H., Schnitzke, K., and Ziegler, G. 1975. *IEEE Trans. Magn.,* **MAG-11,** 421.

Hinds, J. J., and Hartwig, W. H. 1971. *J. Appl. Phys.,* **42,** 170.

Hollenhorst, J. N. 1979. *Phys. Rev. D,* **19,** 1669.

Isagawa, S., Kimura, Y., Kojima, Y., Mitsunobu, S., and Mizumachi, Y. 1974. In *Proc. 9th Int. Conf. High Energy Accelerators.* Stanford, Calif.: SLAC, p. 147.

Kaplun, Z. F. 1974. *Elektronnaya Tekhnika: Elektronika SVCh,* No. 8, p. 3.

Khaikin, M. S. 1961. *Pribory Tekhnika Eksperimenta,* No. 3, 104.

Kikoin, I. K., ed. 1976. *Tables of Physical Quantities.* Moscow: Atomizdat.

Kimura, S., Suzuki, T., and Hirakawa, H. 1981. *Phys. Lett.* **81A,** 302.

Klein, M., Proch, D., and Lengeler, H. 1980. Preprint GHW. Wuppertal, Germany.

Kneisel, P., Stolz, O., and Halbritter, J. 1979. *IEEE Trans. Magn.,* **MAG-15,** 21.

Kneisel, P., Stoltz, O., and Halbritter, J. 1972. In *Proc. Appl. Superconductivity Conf.* (Annapolis, Maryland), IEEE Publ. No. 72 CHO 682-5-TABSC, p. 657.

Kneisel, P., Küpfer, H., Schwarz, W., Stoltz, O., and Halbritter, J. 1977. *IEEE Trans. Magn.,* **MAG-13,** 496.

Kochubei, A. D., and Mitrofanov, V. P. 1978. *Pribory Tekhnika Eksperimenta,* No. 5, 144.

Kolesov, V. V., Panov, V. I., and Petnikov, V. G. 1976. *Pribory Tekhnika Eksperimenta,* No. 4, 169.

Labusch, R. 1965. *Phys. Stat. Sol.,* **10,** 645.

Lagomarsino, V., Manuzio, C., Parodi, R., and Vaccarone, R. 1979. *IEEE Trans. Magn.,* **MAG-15,** 26.

Lamb, W. E. 1965. In *Quantum Optics and Electronics,* ed. C. DeWitt, A. Blandin, and C. Cohen-Tannoudji. New York: Gordon and Breach, pp. 331-381.

Landau, L. D., and Lifshitz, E. M. 1965. *Theory of Elasticity.* Moscow: Nauka.

Landau, L. D., and Lifshitz, E. M. 1959. *Electrodynamics of Continuous Media.* Moscow: Fizmatgiz.

Landau, L. D., and Rumer, Yu. B. 1937. *Sov. Phys.,* **11,** 18.

Lemanov, V. V., and Smolensky, G. A. 1972. *Uspekhi Fiz. Nauk,* **108,** 465.

Libowitz, G., ed. 1976. *Fracture,* vol. 7, part 1.

Maki, K. 1965. *Phys. Rev. Lett.,* **14,** 98.

Malakhov, A. N. 1968. *Fluctuations in Self-Excited Oscillating Systems.* Moscow: Nauka.

Marcuse, D. 1972. *Light Transmission Optics.* New York: Van Nostrand.

Mason, W. P. 1964. *Physical Acoustics Principles and Methods,* v. 1-15. New York: Academic Press.

Mattis, D. C., and Bardeen, J. 1958. *Phys. Rev.,* **111,** 412.

Maxwell, E. 1964. *Progr. Cryog.,* **4,** 123.

McGuigan, D. G., Lam, C. C., Gram, R. Q., Hoffman, A. W., and Douglass, D. H. 1978. *J. Low Temp. Phys.,* **30,** 621.

Mende, F. F., Bondarenko, I. N., and Trubitsyn, A. V. 1976. *Superconducting and Refrigerating Resonance Systems.* Kiev: Nauk Dumka.

Meyer, W. 1977. *IEEE Trans. Microw.,* **MT-25,** 1092.

Meyerhoff, R. W. 1969. *J. Appl. Phys.,* **40,** 2011.

Miller, P. B. 1960. *Phys. Rev.,* **118,** 928.

Minakova, I. I., Nazarov, V. I., Panov, V. I., and Popelnyuk, V. D. 1978. *Pis'ma v. Zh. Tekh. Fiz.*, **4**, 172.

Mittag, K., Hietschold, R., Vetter, J., and Piosczyk, B. 1970. In *Proc. Proton Linear Accelerator Conf.* (Batavia, Illinois), p. 257.

Mittag, K., Schwettman, H. A., Schwarz, H. D., *et al.* 1973. *JEKP,* Note 121, (Kernforschungszentrum Karlszuhe, Germany).

Mitrofanov, V. P., and Frontov, V. N. 1974. *Vestn. Mosk. Univ., Ser. Fiz. Astron.*, No. 4, 478.

Mitrofanov, V. P., and Shiyan, V. S. 1979. *Kristallografiya*, **24**, 307.

Mitrofanov, V. P., Ovodova, L. G., and Shiyan, V. S. 1980. *Fizika Tverdogo Tela*, **22**, 1945.

Nguyen Tuong Viet 1967. *C.R. Acad. Sci.*, **264**, 1227.

Nguyen Tuong Viet, and Biquard, F. 1966. *C.R. Acad. Sci.*, **262**, 590.

Novikov, S. I. 1974. *Thermal Expansion of Solids.* Moscow: Nauka.

Nowick, A. S., and Berry, B. S. 1972. *Anelastic Relaxation in Crystalline Solids.* New York: Academic Press.

Oide, K., Tsubono, K., and Hirakawa, H. 1980. *Japan. J. Appl. Phys.*, **19**, 123.

Padamsee, H., Kirchgessner, J., Tigner, M., *et al.* 1976. Preprint CLNS-340. Cornell University, Ithaca.

Paik, H.-J. 1980, in *Abstracts of 9th Int. Conf. Gen. Rel. Grav.* (Jena, DDR), p. 344.

Panov, V. I. 1980. *Pribory Tekhnika Eksperimenta,* No. 5, 124.

Panov, V. I., and Khalili, F. 1980. In *GR-9 Conf. Abstracts,* p. 397.

Panov, V. I., and Rudenko, V. N. 1979. *Radiotechn. i Electron.*, **24**, 1040.

Panov, V. I., and Sobyanin, A. A. 1984. *LT-A Contributed Papers,* eds. V. Eckern, A. Schmid, W. Weber, and H. Wühl (Elsevier Science Publishers, B. V.), in press.

Panov, V. I., and Sobyanin, A. A. 1982. *Pis'ma Zhur. Eksp. Teor. Fiz.*, **35**, 44.

Passow, C. 1972. *Phys. Rev. Lett.*, **28**, 427.

Pierce, J. M. 1974a. *Meth. Exp. Phys.*, **11**, 541.

Pierce, J. M. 1974b. In *Methods of Experimental Physics,* ed. R.V. Coleman. New York: Academic Press, 541.

Pierce, J. M. 1973. *J. Appl. Phys.*, **44**, 1342.

Pierce, J. M., Schwettman, H. A., Fairbank, W. M., and Wilson, P. B. 1964. In *Proc. Int. Conf. on Low Temp. Phys.* (New York), Part A, p. 396.

Pippard, A. B. 1954. In *Advances in Electronics and Electron Physics,* vol. 6, ed. L. Marton. New York: Academic Press.

Pound, R. V. 1946. *Rev. Sci. Instrum.*, **17**, 490.

Rayleigh, J. W. S. (Baron) 1929. *The Theory of Sound.* New York: Macmillan.

Remis, G. A., ed. 1949. *Technical Measurements at Centimeter Wavelengths,* vol. 1. Moscow: Soviet Radio; translated from the

English.

Reoter, G. E., and Sondheimer, E. H. 1948. *Proc. Roy. Soc. London A.*, **195**, 1042.

Schnitzke, K., Martens, H., Hillenbrand, B., and Diepers, H. 1973. *Phys. Lett. A.*, **45**, 241.

Schwettman, H. A., Turneaure, J. P., Fairbank, W. H., Smith, T. I., McAshan, M. S., Wilson, P. B., and Chambers, E. E. 1967. *IEEE Trans. Nucl. Sci.*, **NS-14**, 336.

Shevchenko, V. V. 1969. *Smooth Transmission in Open Waveguides.* Moscow: Nauka.

Skanavi, G. I. 1949. *Physics of Dielectrics.* Moscow and Leningrad: Gostekhizdat.

Smagin, A. G. 1977. *Pribory Tekhnika Eksperimenta*, No. 1, 166.

Smagin, A. G. 1975. *Pribory Tekhnika Eksperimenta*, No. 6, 157.

Smagin, A. G. 1974. *Pribory Tekhnika Eksperimenta*, No. 6, 143.

Smagin, A. G., and Yaroslavsky, M. I. 1970. *Piezoelectricity of Quartz and Quartz-Crystal Oscillators.* Moscow: Energiya.

Smarr, L., ed. 1979. *Sources of Gravitational Radiation.* Cambridge: Cambridge University Press.

Smirnov, A. I. 1973. *Trudy Vsesoyuzn. Nauchno-issl. in-ta fiziko-tekhn. i radioizmerenii*, **9**, 36.

State Standard of Time and Frequency of the USSR, 1978. "Corrections of carrier frequencies of radio station GBR and RBU," *Measuring Techniques*, 1978, No. 4, 34.

Stein, S. R. 1975. In *Proc. 29th Annual Symp. Frequency Control.* Washington D.C.: Electronic Industries Association, p. 321.

Stein, S. R. 1974. Unpublished Ph.D. thesis, Stanford University.

Stein, S. R., and Turneaure, J. P. 1978. In *Proc. AIP Conf. No. 44* (Charlottesville), p. 192.

Stein, S. R., and Turneaure, J. P. 1976. *Atomic Masses and Fundamental Constants*, eds. J. H. Sanders and A. H. Wapstra. New York: Plenum Press, **5**, 636.

Stein, S. R., and Turneaure, J. P. 1973. In *Proc. 27th Annual Symp. Frequency Control* Washington D.C.: Electronic Industries Association, p. 414.

Stein, S. R., and Turneaure, J. P. 1972. *Electron Lett.*, **8**, 321.

Suzuki, T., Tsubono, K., and Hirakawa, H. 1978. *Phys. Lett. A.*, **67**, 10.

Thorne, K. S. 1980. *Rev. Mod. Phys.*, **52**, 285.

Thorne, K. S., Drever, R. W. P., Caves, C. M., Zimmermann, M., and Sandberg, V.L. 1978. *Phys. Rev. Lett.*, **40**, 667.

Tomikawa, Y. 1979. *IEEE Trans. Sonics Ultrason.*, **SU-26**, 259.

Turneaure, J. P. 1967. Report HELP-507. Stanford, Calif.: Stanford University.

Turneaure, J. P. 1972. In *Proc. Appl. Superconductivity Conf.* (Annapolis, Maryland), IEEE Publ. No. 72 CHO 682-5-TABSC, p. 621.

Turneaure, J. P., and Nguyen Tuong Viet 1970. *Appl. Phys. Lett.*, **16**,

333.

Turneaure, J. P., and Weissman, I. 1968. *J. Appl. Phys.,* **39,** 4417.

Turneaure, J. P., Will, C. M., Farrell, B. F., Mattison, E. M., and Vessot, R. F. C. 1983. *Phys. Rev. D,* **27,** 1705.

Unruh, W. 1979. *Phys. Rev. D,* **19,** 2888.

Unruh, W. 1978. *Phys. Rev. D,* **18,** 1764.

Vasiliev, S. I., and Panov, V. I. 1980. Preprint No. 3/1980. Physics Faculty, Moscow State University.

Veilsteke, A. 1963. *Foundations of Quantum-Mechanical Amplifiers and Generators.* Moscow: IL.

Veitch, P., Blair, D. *et al.* 1985. In *Proc. of the LT-17 (Low Temperature) Conference,* eds. V. Eckem, A. Schmid, and W. Weber, in press.

Vessot, R. F. C., Levine, M. W., and Mattison, E. M. 1977. Preprint No. 895. Harvard/Smithosonian: Center for Astrophysics.

Vinogradov, V. S. 1969. *Trudy FIAN,* **48,** 76.

Vzyatyshev, V. F., and Dobrosmyslov, V. S. 1977. *Trudy MEI,* **341,** 59.

Weber, J. 1961. *General Relativity and Gravitational Waves.* New York: Wiley-Interscience.

Weissman, J., and Turneaure, J. P. 1968. *Appl. Phys. Lett.,* **13,** 390.

Werthamer, N. R. 1969. In *Superconductivity,* ed. R. D. Parks. New York: Marcel-Deccer, vol. 1, ch. 6.

Williams, D. L. 1962. *Proc. Phys. Soc.,* **79,** 594.

Zener, C. M. 1948. *Elasticity and Anelasticity of Metals.* Chicago: University of Chicago Press.

Index

143   Index

CPSIA information can be obtained
at www.ICGtesting.com
Printed in the USA
LVHW08s1025140718
583482LV00001B/6/P